SPACE MARKETING

Space Marketing

A European Perspective

by

Walter A.R. Peeters
*International Space University,
Strasbourg, France*

Space Technology Library

Published Jointly by
Microcosm Press
El Segundo, California

Kluwer Academic Publishers
Dordrecht / Boston / London

Library of Congress Cataloging-in-Publication Data

ISBN 0-7923-6744-8

Published jointly by
Microcosm Press
401 Coral Circle, El Segundo, CA 90245-4622 U.S.A.
and
Kluwer Academic Publishers,
P.O. Box 17, 3300 AA Dordrecht, The Netherlands.

Sold and distributed in North, Central and South America
by Microcosm
401 Coral Circle, El Segundo, CA 90245-4622 U.S.A.
and Kluwer Academic Publishers,
101 Philip Drive, Norwell, MA 02061, U.S.A.

In all other countries, sold and distributed
by Kluwer Academic Publishers,
P.O. Box 322, 3300 AH Dordrecht, The Netherlands.

Printed on acid-free paper

All Rights Reserved
© 2000 Microcosm, Inc. and Kluwer Academic Publishers
No part of the material protected by this copyright notice may be reproduced or
utilized in any form or by any means, electronic or mechanical,
including photocopying, recording or by any information storage and
retrieval system, without written permission from the copyright owner.

Printed in the Netherlands

TABLE OF CONTENTS

Foreword	ix
Preface	xi

1. The Space Market — 1
 1.1 Introduction — 1
 1.2 Global Market Volume — 2
 1.3 Employment in the Space Sector — 9
 1.4 Global Trends — 10
 1.5 Trends in Europe — 12
 1.6 The Changing Environment — 19
 1.7 Commercialisation — 23
 1.8 Financing — 29
 1.9 Public-Private Partnership — 33
 1.10 Conclusion — 42

2. Use of a Space Nonprofit Marketing Mix — 45
 2.1 Introduction — 45
 2.2 Marketing in its Historical Context — 46
 2.3 Definition of the Nonprofit Sector — 51
 2.4 Differences between the Profit and Nonprofit Sector — 56
 2.5 Criticism against Marketing in the Nonprofit Sector — 60
 2.6 The Choice of a Marketing Mix in the Space Sector — 64
 2.7 Conclusion — 72

3. The Space Product — 75
 3.1 Introduction — 75
 3.2 Specificities of Space Products — 76
 3.3 The Product Choice — 82
 3.4 Efficiency in Space Products — 87
 3.5 The Space Product Development Cycle — 91
 3.6 Spin-off — 101
 3.7 Conclusion — 110

4. Price of Space Projects — 113
 4.1 Introduction — 113
 4.2 Price Definition in the Nonprofit Environment — 114
 4.3 Apparent Cost of Space Projects — 116
 4.4 The Problem with Overruns — 118

	4.5	Cost Estimating	122
	4.6	Design to Life Cycle Cost	126
	4.7	Contractual Provisions	130
	4.8	Cost Control and Risk Management	137
	4.9	Space Insurance	143
	4.10	Conclusion	147

5.	**Distribution of Space Products**	**151**
	5.1 Introduction	151
	5.2 Distribution of Services	152
	5.3 Space Data Distribution	154
	5.4 Increased Information Needs : The TLC Concept	158
	5.5 The Innovation Flow	164
	5.6 Spin-off Distribution	168
	5.7 The Use of the Internet	173
	5.8 Intellectual Property Aspects and Distribution	177
	5.9 Conclusion	183

6.	**Promotion of Space Activities**	**187**
	6.1 Introduction	187
	6.2 Promotion versus Communication	188
	6.3 European Space Promotion	192
	6.4 Design of a New Communication Strategy	196
	6.5 A European Space Message	199
	6.6 The Space Audience in Europe	204
	6.7 The Choice of Media	208
	6.8 A Specific Problem : The Importance of Names	213
	6.9 Ethical Aspects of Space Communication	214
	6.10 Conclusion	220

7.	**Space and Philosophy**	**223**
	7.1 Introduction	223
	7.2 Traditional Resistance towards Changes	225
	7.3 The Slow Progress of Space Development	227
	7.4 Changes and Innovation in the Space Sector	230
	7.5 The Underlying Philosophical Dimension	236
	7.6 The Cross-cultural Cooperation in Space Projects	243
	7.7 Conclusion	254

8.	**Case Study: International Space Station Commercialisation**	**257**
	8.1 Introduction	257
	8.2 Short Description of the Programme	259
	8.3 Present Commercialisation Evolution in NASA/CSA	263

	8.4	ISS Commercialisation in Europe	267
	8.5	The ISS Product	270
	8.6	Price of the Station	277
	8.7	Physical Distribution	281
	8.8	Promotion of ISS	284
	8.9	Philosophy and ISS	288
	8.10	Conclusion	292
9.	**Case Study : Space Tourism**	**295**
	9.1	Introduction	295
	9.2	Global Boundary Conditions	296
	9.3	Distinctions in Space Tourism Approaches	298
	9.4	Analysis of the Market	301
	9.5	Products Offered	305
	9.6	Price of Space Tourism	313
	9.7	Physical Distribution and Target Groups	317
	9.8	Promotion for Space Tourism	319
	9.9	The Philosophical Dimension of Space Tourism	320
	9.10	Marketing Plan	323
	9.11	Conclusion	325

List of Acronyms   329

Index   333

FOREWORD

The idea for this book has followed an unusual trajectory. After an initial period of professional activities in industry, I joined ESA, the European Space Agency, in 1983. Based upon my dual background in engineering and economy, I was involved mainly in project control and development management activities.

Upon creation of the European Astronaut Centre, in the ESA Directorate of Manned Space Activities, I got involved in different assignments of ESA astronauts. This way, I had the chance to get better acquainted with major space organisations such as NASA, DLR, CNES and NASA and, in particular, with the Russian space organisations RAKA, ZPK, IBMP and RSC Energia. Besides the more evident cross-cultural differences, it was equally interesting to note how different organisations in different political and economic systems operated. The recruitment of new astronauts and their basic training in Europe brought also forward the aspect of "Space Marketing". Indeed, astronauts are solicited highly in interviews and public events, and are the first ones confronted with the question "what is the purpose of space activities?"

In fact this was the origin of a small course on Space Marketing, in order to provide the astronaut candidates with sufficient and structured information on such questions. Later, the course was extended to postgraduate students, inter alia at the Technical University of Delft (the Netherlands). It was at these courses that the need arose to group the material, which can be found in different excellent handbooks, into one book focusing on this specific subject. Therefore, this book would not have been possible without the various comments and inputs from the astronauts and students who were confronted with the basic texts as course notes.

As it was impossible to combine writing of the book with full-time activities, I am indebted to the ESA management who allowed me a sabbatical period to research further on the topic. This gratitude is directed specifically and my director at ESA, J. Feustel-Büechl and to my direct superior at the time of the request, E. Slachmuylders, but also to the staff of the Personnel department, J.C. Spérissen, C. de Cooker and V. Percheron who provided the necessary support.

For the sabbatical period I chose to go back to the environment where I received my first introduction to Marketing, to the Department of Applied Economics at the University of Louvain (Belgium). I was happy to find myself back in touch with the person who introduced me some 30 years ago to the science of Marketing, Prof. J. Leunis, and who provided me once again with the necessary guidance. My special thanks also go to Prof. P. Vandenabeele, the Dean of the Department, for providing me with excellent facilities, which allowed me to process this text. Specific thanks also to Prof. L. Warlop, who kindly reviewed some chapters in his field of interest and provided me with valuable inputs as well as to Prof. L. Bouckaert for reading the philosophical oriented chapter.

Evidently, the breadth of the topic covered forced me to go into specialised areas far beyond my competence. I was therefore grateful for the advice, material and support I received from ESA colleagues on a number of areas in their field of expertise; without being limited I would like to mention A. Farand (legal aspects), W. Thiebaut (insurance aspects), R. Elaerts (promotion), V. Kayser (commercial law), S. Jähn (Russian aspects), R. Oosterlinck (Intellectual property), U. Christ (communications), R. Belingheri (Risk Management), and V. Damann (space medicine).

To put all the ideas in an acceptable format could not be done without editing advice and help from Dr. R. Henderson, who was so kind to read the whole text and provided many valuable inputs, as well as Dr. H. Blom from Kluwer Publishing.

However, it has to be made clear that all the views expressed in this book are personal, and do not necessarily reflect the views of ESA.

PREFACE

Since 1957, when the first "Sputnik" was launched and nowadays, space activities have undergone a considerable change. Started off as a public endeavour, the politically motivated "Space Race" resulted in rapidly progressing successes. The first man on the moon, only 12 years after the launch of a first satellite, illustrates well this enormous evolution. As a result of geopolitical changes and reduced public funding possibilities, governments have initiated a number of steps from the regulatory and legal viewpoint to encourage industry to take a larger role in space activities.

At present, we are witnessing a transient stage whereby public investments are still considerable but gradually reducing. A number of courageous private initiatives have emerged over the last years but, as can be witnessed in the case of Iridium, risks are still considerable in this new space market.

Many of the present commercial successes are therefore, such as in the case of telecommunications, earth observation and launchers, still based upon initial infrastructure funding with public money. Such public funding will be needed still in a number of novel space areas where the risks are considered too high by industry and private investors.

Another element of importance in this respect is the progress of scientific and technological developments in other fields, such as multimedia and bioengineering. These areas offer equally good commercial perspectives and attract private capital, whereas on the other hand also governments are making more public investment in these promising areas. As such, space activities have to compete with other disciplines for the ever more scarce public resources. Whenever such a competitive element appears in a free economical system, this leads to a form of marketing, formal or informal. Commercial space "business to business" activities have already for years followed the industrial marketing approaches which are well documented in many textbooks on hi-tech marketing.

The objective of this book is therefore to target to a number of different and less obvious marketing aspects, namely

- How can industry market new projects in an environment that is still driven by public investments?
- Is there a way to combine forces between industry and governments to explore new markets?
- And, above all, how can we better "market" space activities to safeguard the necessary level of public funding?

Schemes such as PPP (Public-Private-Partnership) are considered here as an excellent tool to handle this transient phase and will be suggested wherever appropriate. Such approaches are very dependent on what is called in marketing the "macroeconomic environment". Even after the recent changes, political and strategic aspects are still of high importance in space activities which is reflected in deregulation efforts on the one hand and for example export restrictions on the other hand. Also the attitudes of governments in keeping space activities as a public good, or to privatise as much as possible, play a considerable role in the mode with which the market operates. From this point of view, Europe has been basically taken as the pilot environment, mainly for the reason that transition from public to private activities is strongly evolving in Europe with the obvious result of developing new instruments. Such instruments have further been developed in for example the U.S., and therefore many cross-references are given in the text.

This does not mean at all that the text only targets the European reader. Marketing principles are evidently generically applicable to each specific case, irrespective of the environment. Similarly, marketing tools and the construction of an appropriate Marketing Mix are universal instruments and applicable for all cases. The major difference lies in the emphasis on the tools to be used within such a Marketing Mix in order to come to an optimal result and it is in this tuning process that the macroeconomic factors play their role.

These considerations bring us to the subject of target groups for which this book can be of interest, to which one could count:

- **Space professionals in general**, as well in the public as in the private sector.
  Besides the generic interest on "how to market space activities better", the attention has to be drawn to the fact that also the private sector profits directly (via contracts) but also indirectly (via public investments) from public space budgets.

- **Industrial space marketing professionals.**
  The reader will note that a lot of general space market material has been collected and much of it in the form of tables. As one of the basic principles of marketing is information collection, this collection of material in one work can constitute a good reference base.

- **Public space strategists.**
  Reducing resources are forcing public space administrators not only to do better marketing, but also to look for alternative solutions. Financing methods such as the PPP concept or sponsoring could help in this respect, but also the awareness that a number of regulatory and organisational problems will have to be solved soon to allow for implementation.

- **Marketing specialists in general**, and Industrial Marketing specialists in particular.
  A lot of emphasis is put in this area on technology transfer and spin-off processing. In the field of hi-tech markets, the space environment is taking a prominent role in this respect.

- **Post-graduate space students.**
  The core of this book is based upon lectures which were prepared for specialised courses, such as the "Master in Space Systems Engineering Programme" at Delft University (the Netherlands). The need for a more substantial textbook, as voiced by a number of students, was the catalyst for this book.

The structure of the book follows these orientations and covers three blocks:

Block 1: Background Information

This block covers a chapter with detailed information on the space market and some general marketing principles are recalled.

Block 2: Marketing Tools

Classical marketing is normally described as a Marketing Mix of four tools, generally known as the 4Ps: (Product, Price, Promotion and Physical Distribution). The opinion is maintained that space activities cannot be marketed properly with these tools only, and therefore a fifth P, Philosophy, is added in a separate chapter.

Block 3: Implementation / Case Studies.

A choice had to be made because each space application with a commercial potential could fall under this category. The final choice has been based upon following considerations:

- The first case, on **Commercialisation of the International Space Station**, is a realistic one actually taking place at this present period of time.

- The second case, on **Space Tourism**, may sound a bit remote but still is of high interest in the framework of this book. There is a considerable chance that this sector will develop purely on the basis of private investments and as such is an excellent candidate for marketing strategies.

Having these elements in mind, one could tentatively indicate which parts of the book are of interest for the different readers, as per following table:

	GENERAL SPACE INTEREST	INDUSTRY SPECIALIST	PUBLIC SPACE SPECIALIST	MARKETING EXPERT	STUDENTS
Block 1: Space Market Marketing	XXX	XX0	XXX	XX0	XXX
Block 2: Marketing Tools	X	XX	XX	XX	XX
Block 3: Case Studies	X	S	S	S	X

XX	: Recommended
X	: Of interest
0	: No specific need
S	: Depending on specific interest in the topic.

Briefly, the chapters can be summarised as follows:

Chapter 1: The Space Market

In the first instance, economic data on space activities and the outlook thereof are compiled in this chapter. These data provide an insight into the market structure and in future trends. Emphasis is put on the changed environment, specifically in view of reduced public funding and the trend to increasing commercialisation. In view of this different financing schemes are discussed with an emphasis on the PPP-approach.

Chapter 2: Use of a Space Nonprofit Marketing Mix

Based upon the rationale that marketing is fully introduced in the commercial space environment, emphasis is put on the less usual nonprofit marketing. This latter aspect is illustrated with a number of potential criticisms. In the second part, for the readers less familiar with Marketing some essential marketing principles are recalled. In comparison to other fields, a Marketing Mix using five tools (the classical 4Ps and Philosophy) is selected for the space marketing approach.

Chapter 3: The Space Product

To start with, it is explained that it is not the technical product side that is the subject of this chapter but rather the marketing aspects of the product. In similarity with industrial new products, the choice of the right product, in relation to the market requirements, is discussed. Specifically the need for efficiency, in view of potential public criticisms of space activities and reliability are highlighted in the chapter. Moreover, emphasis is put on the spin-off products, which are considered as a paramount marketing instrument.

Chapter 4: Price of Space Projects

The elements as to why the general public considers that space projects are expensive are elaborated. A specific problem in this context are the cost overruns which have shed a negative light on space endeavours. Therefore, the tools to control the costs are highlighted; ranging from a priori cost estimation methods to the appropriate choice of the contract type as a measure against overruns. Special attention is given to an underestimated cost factor, the space insurance cost.

Chapter 5: Distribution of Space Products

A distinction is made between the data received from space activities and the technological products. The change in attitude for the first category, under influence of cheaper communication techniques such as the Internet is discussed. For the second aspect, the role of space technologies in the "Technology Life Cycle" is highlighted. Important in this context are the intellectual property aspects. Present regulations do not facilitate commercial applications and therefore, specifically in this area, much preparatory work still has to be done in the next few years.

Chapter 6 : Promotion of Space Activities

Specifically in the public sector promotion activities must be considered very carefully in order to avoid criticism ("waste of public money"). The messages to be conveyed, the medium used for this and, above all, the target audience need to be chosen with care. Specific marketing problems such as the correct choice of names and the use of fear (asteroids) are discussed in more detail in view of their applicability.

Chapter 7 : Space and Philosophy

After a rapid start, space activities have not developed as quickly as generally expected. The reasons for this are analysed and remedial steps proposed to reuse the underlying philosophical dimension of space activities in this respect. The cross-cultural cooperation in large space projects and the effect of people, with considerably differing backgrounds and working together is highlighted as a testbed for a more global societal structure in the future.

Chapter 8: Case study: The International Space Station Commercialisation

The International Space Station has been financed with public funding and is now proposed to the private community for utilisation. A number of first projects and potential areas are discussed from a Marketing point of view. Even if this intention is clearly expressed, a number of areas still need to be explored before such commercialisation is possible, specifically from the regulatory side. These areas, such as pricing, access rights and Intellectual Property aspects are further detailed in the chapter.

Chapter 9: Case study: Space Tourism

The framework of space tourism is rather different. It can safely be expected that this promising field of commercial space activities will be developed in a purely business approach and supported by the expanding tourism industry. Marketing of this new product will require a dedicated approach, which is suggested in this chapter. The cost for a broader public is still beyond reach and a model is proposed to determine acceptable upload costs. Also in this area, a number of regulatory issues will have to be tackled well in advance the next few years.

Chapter 1

THE SPACE MARKET

1.1 INTRODUCTION

As with most of the past endeavours of mankind to explore the world and beyond, space activities started off as a publicly financed, scientific activity.

Most of the early pioneers were scientists, who invested a lot of personal energy in developing space science and technology. In common with many new discoveries, these scientists encountered a lot of resistance, particularly in academic circles. A well-known story in this respect was the fundamental work of the German space pioneer Oberth, whose doctorate thesis "Die Rakete zu den Planetenräumen" [1] was even rejected in 1923 for being "insufficiently realistic".

As so often with a number of important developments, such as the development of the "Ironclad" ships (the Merimac, in 1863) and the "Heavier-Than-Air Flying Machines", now called airplanes, (bought by the U.S. Signal Corps in 1908), the military were the first people interested in such new developments. Historically, this was already the case in China during the Song dynasty (13th century) when amongst others rocket technology was successfully used in the battle against the Mongols of Genghis Kahn, in 1232 [2].

This led unavoidably to a certain focusing of early developments for military purposes. The first confrontation that the general population had with rocket technology was during the Second World War. The Katyushka "Stalin Organs" in Russia and the V1/V2 in Western Europe may still have left bitter memories and a psychological reluctance to support space activities.

After the Second World War, there was a boom in space activities. The "Space Race" had started, with the political objectives of being the first to put a man in space and a man on the moon. Due to this highly political character, such projects were administered in a world of secrecy; for example the first space related facilities were not even marked on maps in Russia, and, inevitably, there was no transfer to commercial applications.

However, gradually rocket and satellite technology became readily available and scientific satellites were developed and launched. In order to relay data and pictures, instruments and payloads needed to be developed, which soon demonstrated broader application possibilities than the purely scientific. The study of meteorological phenomena has led to weather satellites, the transmission of pictures has led to TV-broadcasting, geological and geographical survey of the Earth has led to a number of Earth Observation applications and, in general, the need to "communicate" data with satellites evolved in the Telecommunications satellites.

With the basic technology available, there was a gradual step to commercial applications. Transponders on scientific satellites were "leased" to commercial firms and scientific pictures were used for roadmaps and land survey purposes. In this way, space transferred from science to business, and now we can validly talk about a Space Market. Like any new market, the Space Market is in a transition stage, where important foundations are still missing (such as the legal framework for commercial space activities). There is still an adherence to the consequences of military applications, translated in export control for technology, and, more important, the risks associated with the market are still considerable. The latter effect makes it risky at present for private companies to make the considerable investment required, in addition to the problem of finding financing on the – risk-avert – capital market.

The Space Market will be described in the next pages, which will make clear that we are dealing indeed with an important, "emerging" market. Transition to full commercialisation will be specifically described, because it is one of the most pertinent aspects.

1.2   GLOBAL MARKET VOLUME

To describe the Space Market, we can distinguish between several sets of statistics, which are often difficult to compare, namely

- **The public market**: the figures are in general more clear, but here some figures are not available or not published, specifically in the area of the

military space expenditure. Therefore, it is necessary to compare various sources.
- **The global market volume**: comprises all amounts, public, industrial and services, which are involved in the space segment, including the operation thereof. Hence, it excludes all space related ground equipment such as private satellite dishes, car navigation systems etc.

The most global overview can be found in the report of the International Space Business Council [3]. Overview figures are presented in table 1.1, as well as the expected market growth.

CATEGORY	1998	2002 (EST.)	GROWTH
Infrastructure	55,471	64,389	3.8%
Applications	33,646	56,397	13.8%
Use of space data & assets	4,648	13,307	30.1%
Support Services	3,827	3,728	-
Total	97,593	137,822	9.0%

Table 1.1: Space turnover (in 1998 MEURO)

The categories considered in this table are:

Infrastructure: includes the Space-based (Satellites, Space Station), the Ground-based (Control facilities, spaceports, ground equipment) as well as the launcher elements.

Applications: refers to the telecommunications area in broadest sense (Fixed and Mobile Satellite services, Direct-to-Home television, Digital Audio Radio Services, VSAT services, Broadband and Internet communications).

Use of space data and assets: Remote sensing data (see also figure 1.1), GPS, but also Microgravity and robotics research.

Support Services: Legal, Consulting, Insurance, Publishing, as well as R&D, Universities.

(photo: ESA)

Fig. 1.1: An Application example: earthquake prediction by radar interferometer

From the figures we also have to underline the growth factor of 9 %. Naisbitt [4] expects tourism to be one of the "boom" sectors for the next decennia, with a growth expectation of some 6.1%. Based on these figures the space market will do some 50 % better, which also means that the growth expectation in space activities is roughly twice the expected value for the world economy. Not all markets will grow at an equal rate, but just as one striking example, we can point to the satellite Internet market, which has grown exponentially over the last three years as shown in figure 1.2 [5].

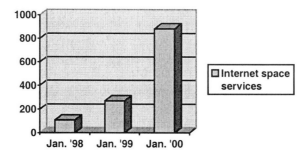

Fig. 1.2: Satellite Internet Market Expansion (in million $)

Let us compare these figures with a second source. Euroconsult is a French consultancy office, having good databases on, specifically European, space activities. Trying to evaluate the competitiveness of European space industry in 1991, they concluded the following [6]

- Present business volume is about ten times smaller in Europe than in the U.S. and two and a half times larger than in Japan
- Expressed in very approximate figures this represents a yearly sales turnover in the U.S. of 25 billion $, in Europe 2.5 billion $ and in Japan 1 billion $
- For the early years of the 21st Century turnovers are expected to be in the range of 55-80 for the U.S., 8-12 for Europe and 3-5 for Japan (all figures in billion $ with a pessimistic-optimistic span)
- For Europe, it is estimated that the industrial turnover is approximately half of the space expenditures at present and this percentage is expected to grow to 65-70%
- Hence, the full market volume is expected to be in the range of 12.5 to 17.5 billion $ in Europe in the early years 21st Century.

Comparison of both data is interesting from a certain perspective. The collected data in a first source [3] are reaching the order of magnitude of the most optimistic forecast data previously assembled in the second source [6]. This is not very surprising in a "booming" market. Products are put on a commercial space market, which could not be envisaged 5 – 10 years before. Older forecasts were more based upon the publicly funded space markets; commercialisation is completely changing the scene. As it is expressed in the preface of the report [3]:

> *While many are unaware of the size and scope of the (space) industry, forecasts show that today's figures will be dwarfed by what is yet to come.*
> *Space is increasingly becoming part of our everyday lives – both in business and at home. Its diverse capabilities range from providing paging, mobile, and remote communications, to television broadcast services, to credit card and ATM bank transaction processing. We use the technology to monitor our planet and provide data products used by business involved with oil and gas exploration, civil planning and site evaluation, disaster management, and the tracking of trucks and cargo. The space industry is no longer limited to the manufacture and launch of spacecraft.*

GLOBAL SPACE EXPENDITURE EXPRESSED IN RELATIVE FIGURES

Another significant indicator in this respect is however the space expenditure in relation to the Gross National Product (GNP). In principle, the GNP can be considered as a yardstick on the ability of countries to invest in space activities, but we note from table 1.2 that the differences are significant. In order to avoid any misunderstandings we have to stress the point that the figures are expressed promille; in other words no country is spending more than 1% of its GNP to space activities in the present timeframe.

COUNTRY	1987	1992	1998
U.S.	5.2	5.2	3.7
Europe	.7	.7	.65
Japan	.45	.5	.7
France	1.1	1.1	1.0
Germany	.5	.4	.4
India	1.2	1.4	.8

Table 1.2: Space Expenditure in function of GNP (expressed as promille)

It is difficult to find comparable data for China and Russia. Figures for 1992 which were presented at symposia were indicating:
- Some 3.5 promille for China
- Some 6 promille for Russia.

Relative figures are more expressive if we compare the absolute ones of for example Germany and France [7]:

In 1997 French public space expenditure amounted to 1859 MEURO, being

- 1092 MEURO for national projects
- 767 MEURO as the ESA contribution.

In 1997, the total figure for Germany was only 697 MEURO, split into:

- 161 MEURO for national projects
- 536 MEURO as the ESA contribution.

One important difference is the military space expenditure, which counted for 773 MEURO in France compared with 53 MEURO in Germany. It goes without saying that France has a considerable number of national notable programs, such as in the field of Earth Observation (SPOT) and military optical observation satellites (HELIOS). As a consequence, France has also developed a much stronger and competitive space industry.

A different approach can be found in Saul [8]. Combining various statistics, he comes to the comparative analysis as per table 1.3. As figures for China and Russia were not available in sufficient detail, the author had to group them under one global factor (i.e. R.O.W., Rest of the World).

COUNTRY	GLOBAL GNP (IN 1997 T$)	PERCENTAGE GNP	PERCENTAGE R&D	PERCENTAGE SPACE
U.S.	7.3	32%	35%	72%
EUROPE	8.7	39%	32%	15%
JAPAN	4.6	20%	23%	6%
R.O.W.	1.9	9%	10%	7%
TOTALS	*22.5*	*100%*	*100%*	*100%*

Table 1.3: Space Expenditure in comparison with R&D and GNP

From this comparison we can note considerable differences: whereas to a certain extent the global R&D effort can be correlated with the GNP, the differences with space efforts are very pronounced. Specifically due to a large effort in the commercial and military space sectors in the U.S., the lower figures for Europe and Japan are apparent.

PUBLIC SPACE EXPENDITURE

Comparison at an Agency level shows a different picture because this depends strongly on the structure of other (e.g. military) expenditure. On the other hand the figures, in general, are easier to trace because we are dealing with public budgets. As an indication, the 1997 figures for major government space budgets are collected in table 1.4. (Euroconsult [6]).

COUNTRY	SPACE BUDGET (1997 MEURO)	AGENCY PART
U.S.	27,406	46.9 % NASA
EUROPE	6,188	60.8 % ESA
JAPAN	2,185	76.3 % NASDA
CIS	710	75 % RAKA
CHINA	360	50% CNSA

Table 1.4: Comparison of 1997 Budgets (in MEURO)

The figures for China and CIS are difficult to evaluate and open to interpretation on what is included and what is not. As an example of this, a recent book on the Chinese space sector [9], p.95 estimates China's 1997 space budget at some 1,380 MEURO, whereas the officially reported figure from the CNSA (Chinese National Space Administration, established in 1993) puts forward a figure of only 250 MEURO.

Another important point that should be mentioned is the synergy between military and civil public budgets in the U.S. In fact, U.S. military space budgets are larger than the public civil ones and, as such, the DoD's space budget is the largest government budget in the world. This has led to the "Dual-Use" concept whereby the R&D for market oriented applications is in fact a spin-off of military developments. Indeed, as the military sector develops and operates these systems, they provide end-to-end testing for new technologies. It goes without saying that this is a considerable advantage for the U.S. companies who operate both in the military and the commercial market. This effect is amplified by the fact that space companies have merged in the U.S. to a considerable extent. Examples of such projects are:

- The Milstar military telecommunications system (25.3 billion $ from 1983-2002)
- Navstar/GPS navigation systems (25 billion $ investments in the period 1974-2016)
- SBIRS Infra-Red early warning satellites (estimated at 22 billion $).

We have to mention here that public space funds, after a period of relative stability in the early 1990's, have decreased considerably. A number of European countries, which were relatively strongly affected by this, are shown in table 1.5. [7]. The figures are actual ones; if an inflation correction factor is added the differences would be even greater.

YEAR	GERMANY	ITALY	U.K.
1992	846	525	226
1993	875	544	242
1994	794	456	214
1995	794	500	223
1996	722	344	210
1997	697	435	210

Table 1.5: Public space expenditure evolution (in MEURO)

1.3 EMPLOYMENT IN THE SPACE SECTOR

One-shot developments consist of a considerable labour portion. In fact, cost estimates are based merely upon engineering man-years including some additional margin for materials (see a.o. Parametric Cost models in chapter 4, such as Koelle's Transcost model). Depending on the sources consulted, however, we find a wide variety of figures. This is due to:

- The non-disclosure of figures in certain countries
- The mix between civil and military staff
- The accounting method, i.e. to which level are people attributed to space activities (Subsystem level? Equipment levels or even component level?)
- The inclusion of associated professions (insurance sector, space education, space camps...).

Based upon a variety of sources, the International Space Business Council comes to a total figure of nearly 1,100,000 persons for 1998 [3]. It has to be admitted that such figures are "enhanced" by the very high employment rates reported in a number of countries where labour costs are very low. A survey of the Chinese space industry [9] reports that some 200,000 people are involved in space activities in China, whereby the CNSA (Chinese National Space Administration) can make use of a large number of technical workers and engineers from the Army. In fact it is reported that, in order to boost this sector, a considerable number of top graduates from military academies were recently transferred to the space sector. Similarly, employment figures in CIS are estimated at some 300,000 persons. For Europe, a complete and detailed survey by Eurospace in 1995 resulted in the breakdown, as shown in table 1.6.

SECTOR	EMPLOYEES (1995)	PERCENTAGE	ESTIMATE (1983)
Space companies (include. services)	33,000	76	16,000
Institutions (ESA/national)	5,500	13	5,000
Universities/ laboratories	5,000	11	4,000
Totals	43,500	100	31,000

Table 1.6: Employment in European Space Activities (1995)

A follow-up of the survey [10] led to a number of additional conclusions:

- The proportion of highly qualified staff in space industry is very high (40% university level)
- The level of production technicians is only 22% in Europe
- Recruitment of experienced staff was considered as a major difficulty by industry
- Recruiting new staff, at a rate of 10% yearly, was considered by industry as very dangerous for stability of the company effectiveness (coaching of new people, training...)
- Major European space companies often employ "high-potential" staff in the space department at first, specifically because of the high quality control standards, in order to move them to other departments afterwards
- With a ratio of 1:8 for European space specialists versus U.S. ones, it was considered that qualified space engineers and technicians were becoming a scarce commodity in Europe.

Eurospace constantly updates the industrial figure, based upon prime data from the Space companies. After an initial slight increase after 1995 the industry figure was approximately 35,000 in 1998 (stabilisation mainly due to restructurisation and mergers in the space sector). A breakdown for some European countries of the 1998 figures [11], is presented in table 1.7.

COUNTRY	EMPLOYMENT	TURNOVER/PERSON (KEURO/YR.)
France	13,252	155
Germany	5,963	172
Italy	5,741	162
U.K.	3,577	137

Table 1.7: Space industry turnover figures for 1998 (major European countries)

One major source of the differences is the fact that French industry has a higher involvement in industrial activities, working under competitive market conditions and more production type of activities.

1.4 GLOBAL TRENDS

In the rapid-changing high-technology environment it is very difficult to make reliable trend-analysis. We can illustrate this with an older article, from 1979 wherein Woodcock [12] after comparative studies and calculations comes up with the following possible space applications with commercial potential.

- Space Communications
- Nuclear Waste Disposal
- Manufacturing in Space
- Space Solar Power.

A more recent grouping by Ashford, in 1991 [13] is already more extensive:

- Sub-orbital transport
- Space tourism
- Solar Power
- Manufacturing in space
- Mining
- Colonisation.

ESA performed a study in 1999 [14], evaluating a number of potential candidates, based upon Feasibility, Affordability, Potential benefit and Spin-off. In a second step, Europe's interest were added to the evaluation process which led to four remaining candidates:

- Mars exploration
- Moon exploration
- Space Solar Power
- Space Tourism.

One shall not forget in this context that mainly public interests, i.e. expansion of the "Space Frontier", were considered as a driving factor. Nevertheless, whereas the first two candidates have a purely scientific interest, the latter two, after initial investments, could equally develop in industrial activities. More business-oriented forecasts are found for example in [3] p.54. In order of ranking the authors are evidently more focusing on business opportunities as follows:

- Tourism
- Solar Power Stations
- Tele-operated Satellite Repair Robots
- Satellite & Space Transfer Services

- Space Business Parks
- Industrial Platforms
- Asteroid Mining
- Lunar & Mars Research Stations.

A tendency that is changing the scene considerably is the entry of commercial entities. These entities bring in products where a governmental organisation, which has less direct contact with the market, would either not even think of, or would not feel comfortable in developing.

Space Burials is just one example of this latter effect. Since 1996 the company Celestis has been selling a service to put (part of) the cremated remains in a probe, and place them in LEO. The company launches in combination with commercial payloads and in December 1999, a Taurus rocket was used whilst launching two Earth Observation satellites. Three launches have taken place up to now, and the plan is to make two further launches in 2000-2001. The Payload is in the order of one kilogram, and customers pay 4,800 $ for each "Burial" (comprising the order of one gram of ashes only…). It is interesting to note that this idea has been very well received in the Asian Market [15], starting in Hong-Kong with plans to expand across China. Also a Japanese agent through a company called Sekise, entered the market in 1999.

1.5 TRENDS IN EUROPE

The European Commission (EC) published the report "Crossroads in Space" [16] in 1991 and concluded that:

- The EC needs a coherent long-term strategy for its space-related activities
- Space for the environment is a priority
- An EC contribution in the field of telecommunications should be continued towards liberalisation
- The EC should give support to the export of European space products
- A more systematic search for synergy between European space and other technologies is needed
- A framework for R&D on re-usable launchers should be established
- The EC role in training the required scientists and engineers has to be strengthened

- The growing cooperation between the EC and ESA should be extended to other parties
- The role of space as a vehicle for wider international cooperation should be encouraged
- Assistance to developing countries to exploit space techniques has to be pursued.

This clearly indicates the willingness of the EC to step into space activities. This will improve liberalisation in legislation and certainly reinforce the role of Europe in this area. Evidently, in line with other steps taken by the EC, one can expect as well a strong trend towards commercialisation and industrial activities. In line with this, from 1997 onwards ESA has changed its philosophy and worked out a strategy based upon four "axes":

1. **Pursuit of scientific knowledge** (stable axis)
2. **Dedication to enhancement of the quality of life** (applied communications, weather forecasting and natural risk management)
3. **Strong European capability** as the key to successful cooperation and competition on the global market (Independent access to Space, GNSS, Manned activities, ...)
4. **Promotion of European Industry**, specifically innovative and added value services (market openings).

These points are based also upon the rationale that the European market share has demonstrated an overall positive trend, as illustrated in table 1.8 (source: Euroconsult 1997)

SEGMENT	1970-1979	1980-1989	1990-1999
Communication Satellites	0 %	25.2 %	28.1%
Launchers	0 %	31.8 %	41.4 %
Meteorological Satellites	5.9 %	23.8 %	36.0 %
Civil Remote Sensing	0 %	42.6 %	33.8 %

Table 1.8: Evolution in European Market share

The space market which is considered, for the next decade, to be "accessible" to Europe is assumed to be composed of:
- Communications: 50 billion $
- Navigation: 1 billion $ (with a ground segment/services of 60 billion $)
- Earth Observation: 10 billion $
- Launchers: 30 billion $.

Again, assuming that Europe can obtain one third of this market, a potential 30 billion $ market is attainable in the next 10 years. These considerations have led to a re-evaluation process in 1999, whereby a long-term visionary group developed a number of "axes", also involving closer synergy with the European Commission for this new orientation [17]. The proposal of this group includes:

1. **Develop European Capacity**
   - Europe to be partner in "Earth-like" planet exploration
   - Europe to invest in launch cost reducing programmes
   - Europe to fully participate in the International Space Station (ISS)
   - Europe to invest in its own navigation systems
   - Europe to maintain observation satellites for security and safety purposes
   - European coordination in orbital frequencies
   - Involvement of European SME's
   - Invest in micro-miniaturisation.

2. **Management of our planet**
   - Participate in programmes for climatological and pollution control
   - Participate in Earthwatch programmes for better prediction of earthquakes and volcanic explosions
   - Early warning systems for climatic changes
   - Better prediction and active "Space Debris" research
   - Evaluation of the risk of impact of large objects.

3. **Initiate further steps**
   - European participation in moon robot programmes
   - Mobilise world-wide activity for a permanent International Moon Station
   - Study the uses of solar energy and astronaut resources
   - Investigate control mechanisms to influence the weather.

4. **In support of this**
   - Improve European Space Education
   - Increase Public Relation initiatives
   - Establish a European Institute for Space Policy.

It has been realised that the traditional way in which ESA does business is not fully compatible with this new environment. Two present developments are noted:
- **Public-private Partnership**, abbreviated with a TLA (Three-Letter Acronym) as PPP, is of generic interest of the space market, and will be described separately further in this chapter.
- **Network of Technical Centres**:

Public expenditure in space is gradually decreasing when expressed in constant terms. This strongly influences the budgets of ESA, but also inter alia those of CNES, DLR, and ASI. As a result, the Organisations have been working together to see how these shrinking budgets can be better used. Competence mapping has revealed that there is a considerable degree of overlap of activities and resources. A number of common fields of activity have been selected and in a pilot phase lasting until 2001, they will be studied jointly to see how these can be used more optimally. Examples are:
- Centralisation of Control Centres activities to specialisations (manned operations in DLR, satellite operations in ESA, launcher operations in CNES)
- Optimal use of tracking stations
- Establishment of one single European Astronaut Corps
- Mutual exchange of specialists in Project Reviews
- Specialisation of space testing facilities.

In marketing terminology, this approach is called the forming of "Strategic Alliances", which can take the form of formal joint ventures or consortia, or, less formal, collaboration on the basis of arrangements. The main reasons and examples of such strategic alliances are adapted from [18] p.255 and presented, as far as applicable to space activities, in table 1.9:

REASONS	INDUSTRIAL EXAMPLES
Access to new capital and activities	Airbus consortium
Greater technical critical mass	U.S. microchip manufacturers alliance
Shared risk and liability	GEC-Alsthom joint venture
Relationship with strategic partners	Airbus consortium
Technology transfer benefits	VW and Bosch Alliance
Reduce R&D costs	GEC / Siemens telecom joint venture
Access to technology	IBM cooperation with Apple
Standardisation	Telecom WAP and UMTS alliances
Management training	Rover / Honda alliance in the 90's

Table 1.9: Reasons for Strategic Alliances

Although the previous considerations have focused on the industrial and technological objectives, one cannot ignore the influence of the "environment" on space activities. Any marketing activity has to face an internal and an external environment. Whereas the internal environment will teach one what is desired, the external environment will tell one what is possible.

This effect is even more conspicuous in activities where the market economy has not established a competitive structure or where legislation is still under development (such as biotechnology, pharmaceutics, e-commerce and also space activities). Treaties or regulations can virtually change concepts overnight and suddenly create new opportunities or destroy other ones. Greenberg [19] has studied extensively the role of Government influences in the U.S. and points out that a Government can influence a large number of uncertainty factors, amongst others by

- reducing the financial risk (e.g. participation in R&D, availability (free of charge) of infrastructure, guaranteed government contracts)
- easing capital formation (e.g. loan guarantees, tax credits, depreciation rules, revenue subsidisation)
- creating new opportunities (e.g. development of new infrastructure, partnership. Legal deregulation)
- supporting awareness and distribution (e.g. international agreements, trade agreements, export and import rules, patents).

In Europe, space is an additional element for political stability. Joint space projects are catalysts for broader inner-European cooperation (as an example, the ECU – European Currency Unit – has been used in European space business long before the introduction of the EURO), and for cooperation with the former Eastern Block-countries (from which some have solicited membership of ESA).

Considering the effects of such a "Borderless Europe", Bonnet concludes [20] p.135:

ESA must also adapt itself to the evolving situation of today's Europe, watch carefully how its main partners will reorient their partnerships in the context of this changing world, and openly and constructively face the challenges posed by the increasing importance of countries like Japan and China.

The role of the agencies in such a scenario will certainly be of a preparatory and implementation nature. The above quoted political guidelines will certainly prevail in planning the future directions to be followed. This "fact-of-life" is probably best summarised by ESA's former Director General, R. Gibson, by quoting from [20] p. 125,

Space is too important to be left to space agencies

We cannot ignore the increasing globalisation factor and its important effects. Cohendet [21] p.53 presents the industrial economical evolution of the of the 20th Century as per table 1.10 in three phases:

	INTERNATIONALISATION 1900-1950	MULTINATIONALISATION 1950-1980	GLOBALISATION 1980-
Economic Environment	Stable	Risky	Uncertain
Nature of the product	Long Life Cycle	Medium Life Cycle	Short Life Cycle
Focus of competition	Quantity and price	Differentiation	Quality
Organisational structures	Hierarchical Headquarters	Hierarchical with subsidiaries	Network type
Nature of the market	Target : Home Country	Target : Local adaptations	Target : Global
Government attitudes	Prevent monopolies	Favour "national champions"	Deregulation

Table 1.10: Three phases of industrial evolution during the 20th Century

A number of the effects we are witnessing in the space sector reflect this globalisation concept. We are seeing further that industry, driven by competitive motives, has started already to concentrate on "network type" of structures, also to target more the global telecommunication and other space application markets. Also, as previously mentioned, the Agencies are leaning more towards a network structure with increased sharing of resources.

The question can, and has often been asked, if this will lead eventually to one single World Space Agency. Some main elements in favour of this are [22] p.26:
- Increasing the global priority given to space
- Raising the level of the world-wide space budget
- Setting space priorities on a global basis

- Creating global standards and cost reduction
- Stressing the peaceful use of space technology.

However, there are a number of important obstacles against this approach:

- Space activities are part of other goals, not ends in themselves
- There are powerful security and commercial interests
- There is limited evidence that such an approach will be effective
- It is questionable that the main space nations would cede control over space activities to such a global body.

But, even if there will not be such World Space Agency (in the near or far future), it is certain that the trend towards organising specific aspects of the world's space activities on an international basis will continue. From this perspective, the ISS is increasingly considered as an excellent testbed for this trend.

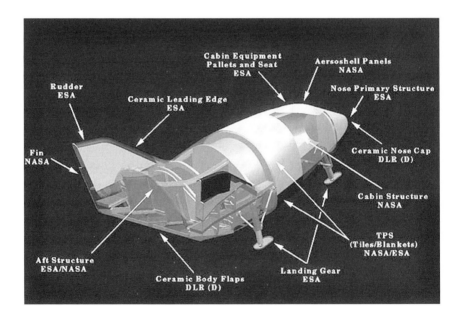

(Photo :ESA)

Fig. 1.3 : The X-38 project : an example of a joint interagency project

1.6 THE CHANGING ENVIRONMENT

REDUCED PUBLIC SUPPORT FOR SPACE ACTIVITIES

On 25 May 1961, only six weeks after Yuri Gagarin's successful flight, U.S. President John F. Kennedy gave the message to Congress:

I believe that the aim of this nation should be to land a man on the moon and return him safely to Earth before the end of the decade

Thus the Apollo program was born and on 16 July 1969, only some 8 years later, Neil Armstrong made man's first step on the Moon.

When he announced the project, there were no lengthy discussions on the budget. Officially in 1969 the Apollo project had a Cost-to-Completion of some 19 billion $, which would represent in present economic terms some 100 billion $, being of the same order of magnitude as the total International Space Station. (The number of ISS budgetary discussions in the U.S. Senate, however, is probably two orders of magnitude greater...).

Nurtured by governments which were striving for political challenges and success, a number of remarkable successes were obtained in the early phase of space exploration. The time between the first Sputnik, the first man in space and the first man on the Moon were so short that expectations were placed very high. In 1967 Wernher Von Braun estimated in his book "Space Frontier" that the first manned mission to Mars would take place around 1986 [23]. After a number of geopolitical changes, which have taken place in the intervening period, the public cannot be motivated anymore in the same way. These changes have led to different priorities, which are influenced by factors such as economical recession and unemployment.

An additional effect is the competition with other areas. In Marketing the term "**Market Myopia**" is used, which means that big, mono- or oligopolistic, Organisations do not "see" that there is competition (as was the case for postal services, public transport, telecommunications, TV) and do not react to this timely. Space activities are now competing for public interest with for example, ecology, biotechnology and informatics. The overall effect explains the shifts in previous table 1.2. To illustrate how considerable this shift has been over the last few years, one should compare the present space expenditure of some 0.5 % of the GNP in the U.S., with the approximately 4 % spending during the Apollo years.

INCREASED COMMERCIALISATION

A new effect has taken place in this period. End-users have understood the economic potentialities of the space "business" and have entered the market. Their objectives are to commercialise a final product. Significant commercial space markets in the short-term can be identified in areas such as

- Direct-to-home-TV
- Multimedia
- Mobile communication
- Navigation services.

Examples in the mobile communication sector in particular are presented in table 1.11 [24]. Most services are presently designed using Medium Earth Orbit satellites, with the exception of Iridium and Teledisc. The commercial failing of the Iridium concept is to a large extent attributable to technical and strategic considerations that were not based sufficiently on a sound marketing concept (handsets too big and communication unit cost too high).

NAME	NUMBER SATELLITES	COST (BILLION $)	OPERATIONAL	PRIMARY SCOPE	COMPANY
Iridium	66 (LEO)	5	1999	Voice, data, fax	Motorola
Globalstar	56	2.6	2000	Voice, data, fax	Consortium
ICO	10	4.7	2000	Voice, fax	Consortium
Skybridge	48	4.2	2001	Multimedia	Alcatel
Teledisc	288 (LEO)	9	2003	Broadband Internet	Motorola

Table 1.11 Planned communication systems

It has to be underlined that, besides the portable telephones, this market also covers e.g. :

- More than 35 million people worldwide receiving their TV via satellite
- The digital audio radio market, starting off in 2001.

Besides the communications area, one must mention also commercial services in the areas of:

- **Commercial launchers**, for example Arianespace ranked in 1999 at eleventh position amongst the world's largest space companies. Others include inter alia Starsem (using Soyuz rockets) and Eurockot (using the Rockot launcher) as well as United Space Alliance in the U.S., ranked in 1999 at position seven.

- **Commercial Spaceports**, such as Spaceport Florida, Wallops Island in Virginia and SeaLaunch activities.

The market volume for both activities together is estimated at some 35 billion $ for the next 10 years. This means that there is a gradual shift from public funding to commercial funding, the effect, thereof, being shown in Fig. 1.4.

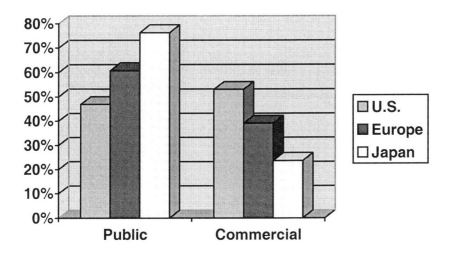

Fig. 1.4: Public vs. Commercial space budgets in 1997 (in %)

Using comparable figures [25], the growth of commercial customers in Europe is rapidly evolving as we can note from fig. 1.5.

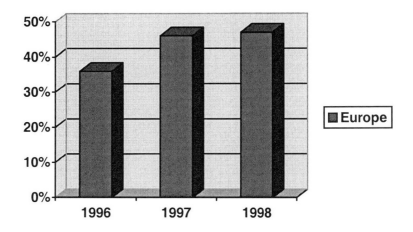

Fig. 1.5 Growth in commercial space budget in Europe

REDUCED PUBLIC EXPENDITURE

To assess the reduction in public expenditure, let us first examine the evolution of public funding over the last few, critical, years. Table 1.12 shows us that, if we express the figures in present purchasing power, than:

- The budgets in Europe have decreased, over 4 years, by more than 18%
- In the U.S., such a decrease has even been in the order of 21%
- Only budgets in Japan still have an increasing trend in the observed timeframe.

YEAR	EUROPE	U.S.	JAPAN
1994	5,498	28,226	1,823
1995	5,309	27,174	2,018
1996	5,120	22,957	2,115
1997	4,583	22,154	2,165

Table 1.12: Public Space Budgets (in 1997 MEURO), source: [3]

Note that it is difficult to compare Russian figures with the ones presented in table 1.12. It can be estimated that the total budget has remained virtually static in this period (around 2000 MEURO), therefore, also de facto strongly decreasing due to inflation. One report [28], when making corrections for exchange rate fluctuations and inflation comes to the conclusion that the Russian civil space budget had decreased by 40 %, compared to 1991.

There is an interrelation between the reduced public expenditure and increased commercialisation. Faced with budgetary problems and aware of the commercial market, governments tend now to turn towards industry by asking them to take a higher risk participation in space projects. It may be of interest to learn how other nonprofit organisations, faced with similar problems, cope with reduced public funding. Keesling and Kaynama [27] have tried to analyse the views of organisations via a questionnaire. They tested three strategies:

- **New Revenue Strategies**, representing initiatives to generate new sources of funding from targeted donors. Within this strategy, "Changing the way services are provided", "Increase service fees" and "Approach new sources" received the highest preference

- **Legitimating Strategies**, which can be considered as an increased promotional activity. The main options preferred in this category are "Collaborative ventures" and "Seek for prominent endorsements"

- **Retrenchment Strategies**, involving efforts to reduce internal costs. This group of strategies was not retained by the respondents, on the basis that they would worsen the situation, probably increase donor scepticism and finally lead to further decreased funding.

Probably due to differences in governance, we have seen that most space Agencies have followed (or have been forced to follow) the latter strategy during the last decade.

1.7 COMMERCIALISATION

The Commercialisation aspect is presently so important that it deserves special attention. More and more industrial firms are entering space markets and also provide end-to-end services (without making use of public services). Although investment is considerable, it is justified if one considers that presently:

- Less than 20% of the world's households have a TV-set
- Less than 5% of the world's population have a personal computer
- Almost half of the world's population lives two hours or more from the nearest phone.

Contrary to Europe, U.S. industry has been preparing itself for this commercial market by mergers. From the 20 major U.S. space companies in 1980, only 3 were left by 1997! (Lockheed-Martin, Boeing and Raytheon). The beginning of this effect was a verticalisation approach, whereby the spacecraft manufacturers expanded into the operator market or the launcher market, in order to create independence. Some early examples of this are:

- Lockheed acquired Marietta (Atlas and Titan launchers) as well as GE Astro Space (Comsats)
- Boeing acquired Rockwell (Delta, GPS, Shuttle) and expanded with Sea Launch and Teledisc.

The perceived advantages are mainly:

- The ability to support turn key services
- More competitive position
- Better feedback from market requirements.

In a second step, the major companies, using mergers and take-overs, consolidated their position in the U.S., basically leading as of now to the three giant companies mentioned above. The advantages of such conglomerates are:

- The possibility to offer "package deals"
- Economies of scale, mainly in production
- Risk investment possibilities (e.g. Boeing have invested 600 million $ in Sea Launch)

The creation of Astrium in Europe goes in the same direction (the final target being to arrive at a second major European space company with some 10,000 employees and 2.7 billion $ annual turnover). This was preceded by the 1998 merger of Alcatel, Aerospatiale and Thomson CSF [30]. Astrium will group the resources of Matra-Marconi, DASA and CASA, eventually also those from Alenia, taking fifth place in the world ranking of space manufacturers [28]. For completeness, it has to be added that Arianespace is ranked at number 11 on the list (1.268 million $ turnover, an increase of 16 % compared to 1997)

RANK	COMPANY	COUNTRY	1998 SPACE SALES
1	Lockheed-Martin Corp.	U.S.	6,711
2	Boeing	U.S.	6,139
3	Hughes	U.S.	5,960
4	TRW	U.S.	4,865
5	Alcatel Space	France	1,851
6	Aerospatiale Matra (*)	France	1,722
8	DASA (*)	Germany	1,348
17	Alenia Aerospazio (*)	Italy	620

(*): Earmarked Astrium partners

Table 1.13: Major Space companies (1998, turnover in million $)

The next step emerging is globalisation, whereby transatlantic alliances are created to enter for the worldwide market. Examples are:

- Alcatel (F), Loral (U.S.) and NPO-PM (Russia)
- Starsem : Aerospatiale and Arianespace (F) with RAKA and Progress (Russia)
- OHB (D) with Fiat-Avio (I) and Yuzhnoe (Ukraine).

The previously mentioned Sea Launch project is one of the most striking examples, because cooperation has led to an innovative concept. This is a self-propelled launching complex (see also figure 1.6) using a modernised drilling platform and providing a floating assembly-testing complex, a storage place for rockets and satellites, as well as a hotel and a flight control centre. The base of the complex is at Long Beach in the U.S., close to the main satellite manufacturer plants.

Table 1.14 summarises the composition of the consortium and the respective involvement. The business model is based upon 6-7 launches per year, with a potential to increase to 11 after 2001.

COMPANY	PART.	COUNTRY	MAIN CONTRIBUTION
Boeing	40%	U.S.	Satellite manufacturing, system
RSC "Energia"	25%	CIS	Launcher technology, mission control
Kvaerner	20%	Norway	Sea Platform, Infrastructure
Yuznoe	15%	Ukraine	Rocket technology, launch pad

Table 1.14: Sea Launch Consortium Composition

(By courtesy of Sea Launch Cons.)

Fig. 1.6: Sea Launch activities

Launches of this type also involve a new aspect on Liability Risk Management due to the entry of private entities in this market. An overview [29]

points out that there are no less than four "layers" of liability which have to be taken into account in such a case, namely:
- The UN Space Treaties and general international law
- The domestic launch regulations/arrangements
- The general domestic law of liability
- The contractual practices of industry.

Conflicting language and the fact that the private entity role, as launch participant, is not fully covered leads to an environment that lacks legal predictability and uniformity and, hence, hinders proper liability risk management. A study [29] on this topic proposes a number of improvements to the UN Space Treaties in order to remedy the situation (e.g. through a dedicated UN Supplementary Protocol to the Space Treaties) as well as improved contractual practices (e.g. by developing "Incoterms" for launch contracts).

In terms of U.S./Russian cooperation one can mention further:

- A joint venture between Lockheed Martin and Khrunichev for the construction of launch boosters (U.S. investment over 100 million $)
- A United Technologies (UTC) and Energomash joint venture for the production of a new booster rocket engine, the RD-180, with a 125 million $ investment. Under this latter operation, UTC directly employs more than 10,000 Russians, making it a considerable western employer in Russia.

This type of globalisation can be explained easily. The emerging markets are, for the next decades, in the Asia-Pacific region. Therefore end-to-end providers will try to team up with operators in these areas in order to enter these markets (For example: the ASTRA – AsiaSat merger in 1998, the EurasSpace Joint Venture between DASA and the China Aerospace Corporation). Analysts expect the Asia-Pacific area to offer a huge market for satellite services due to

- Many low population regions
- Distances that are too expensive to cover by terrestrial lines
- Billions of people wishing to acquire basic telecommunication services
- A need for readily available products, that do not involve development cycles.

The major drive for globalisation are generally grouped in three categories [30]:

1. **Globalising for Internal Efficiency**:
   By coordinating operations, companies reduce costs and become more competitive. Up-front research and high development costs cannot be covered by one or a few markets only; companies become global out of the need to gain more volume.

2. **Globalising to Compete in Homogeneous Markets**:
   Levitt [31] encouraged companies to pursue globalisation by looking at the similarities of markets as opposed to differences.

3. **Globalising for Added Synergies**:
   Leveraging strong positions to help weaker markets to develop is in contradiction to the cost-centre philosophy. However, it allows companies to maximise profits for the entire system and switch back to the weaker markets when they eventually become profitable.

It is evident, with decreasing public funding, that space companies see a benefit in adapting these principles for the upcoming "space market competition".

Although a law has been approved officially in the U.S., called the "Commercial Space Act of 1998" (Public Law 105-303), this only puts up a framework "to encourage the development of a commercial industry". Implementation, such as compliance with International Economic Law and the WTO (World Trade Organization), are now in process. An overview of the present legal problems can be found in [32]. From a legal point of view, some of the main problem areas can be summarised as follows:

- The current regulatory framework for launch activities involves a number of new players having different perspectives
- There is a general lack of regulation of the international space market and, therefore, a need to create a better environment and easier access to the market for the private sector
- Effective licensing procedures for commercial ventures have to be established
- The need for a balance between the improvement of access for private investors in space and the need for safety standards is of great importance.

1.8 FINANCING

Systematically, we can consider the following main sources of financing for a space project [33]:

1. **Venture Capital**: Direct investment in new business by private investors, banks, or investment funds companies. In general the amounts are limited (5-10 million $) and provided in various stages, depending the apparent success rate.
2. **Institutional debt market**: Large banks or Organisations are able to pre-finance large projects. This is mainly applicable in the case of activities resulting from such investments that allow payback of the loan. As an example, Australia's telecommunication services were pre-financed this way.
3. **Publicly owned debt and equity**: Shares are sold to the public to finance commercial activities. Although large sums can be obtained, this is mainly valid for "proven" companies. Such examples are Ford Aerospace and Hughes Communications Inc.
4. **Customer Progress payment**: This is the principal financing form used by governments for the procurement of large system.
5. **Project Finance**: The banks, on the basis of the expected revenues, finance investment. The bank gets a share of the revenue once the system is operational. Although often used e.g. in the oil industry, banks have shown a reluctance to enter such schemes for space projects.
6. **Limited partnerships**: Many associated forms are grouped under this scheme (Joint Venture, R&D partnership) and mainly involve a governmental or institutional partner, sharing the risks. An example in this category is Arianespace.

Looking from the investor's point of view, Halgouet [34] described some of the risks as follows:

- Besides the "normal", technical and market risks, the political risks are considered important (e.g. failure to get licenses, export problems) as well as other legal uncertainties (e.g. intellectual property rights, failure to find insurance)
- Financing risks which are difficult to control due to the rapidly changing market and potential financial claims from operators in case of failures
- Legal and tax risks, due to the new products taxation rules could change, as well as export rules.

We note a number of risk factors which are not emphasised in other projects such as drilling platforms and oil/gas processing, even if the investment sums associated to these projects is much higher most of the time. These areas have been in existence for years and the "environment" has stabilised. A decision to, for example, protect sensitive technologies, or a change in export regulations can appear in the space market to an unforeseen extent. Classical investors do not like to work with unforeseen parameters, and insurance against risk is consequently high, if at all available. All these risk factors lead to the fact that the classical investor is reluctant to invest in space projects, certainly for as long as sufficient terrestrial opportunities for investments are available. Only for "proven" products as telecommunications and TV-broadcasting, are there available reasonable chances for investment sources.

In addition, the technical prerequisites are not fully satisfactory. O'Dale [35] made a survey on existing commercial space activities, together with an outlook for further activities. He considers two main factors as major problems:

- The cost of access to space
- The lack of definable markets.

Access to space will require public investment initially, even if there is industrial interest. Ideally industry would like to have a space transportation vehicle which:

- Has similar capabilities as the Space Shuttle (in terms of manoeuvrability, capacity, communication means)
- Has the reliability and safety of present airliners
- Has an upload cost of less than 1000 $/Kg.

Therefore, the problem is a "Catch-22" one; there is no demand for a low-cost space transportation system without a defined market, but a space market will not emerge without a low-cost transportation system.

One exception to these considerations is the previously mentioned "Venture Capital" approach, specifically when private investors are involved. In this case, equity capital is provided, together with "Value Adding" support, helping young entrepreneurs with management advice at the very early stages of the company's creation. The associated risks are known but by spreading them over a number of promising and carefully selected projects, an overall profit is made.

This approach is typical for the high-technology sector; most Silicon-Valley-type of enterprises started this way. The Venture Capitalist is aware of the risks associated but also of the "high risk, high profit" possibilities. He will examine carefully the various proposals and select a number of promising ones, providing them with a first, modest, financial support (also called "seed-capital"). In case of prospective development, more financing will be provided in the early development stage of the company, in exchange for a portion of the shares. This process takes in total between, on average, 3-7 years, in which period the company will become mature. At some point in that period, the Venture Capitalists will in general exit the company because he is primarily capital gains-oriented. Such exit may run via an IPO (Initial Public Offering on the stock market), a trade sale (to strategic investors such as banks and insurance companies) or to the owners of the company (Buy-Back).

The most well-known Venture Capital firms operating in the space sector are, on the one hand, SpaceVest in the U.S., which launched its first funding in 1995 and managed in 1999 appr. 250 MEURO (in 12 space-related companies) and, on the other hand, GENES in Germany [36].

Also CNES, in France, is taking an initiative to finance high-tech start-up companies, by providing "seed-capital" to companies with technology that could find early commercial applications. CNES would restrict itself to a maximum of 25% share of equity. One of the early projects reported [37] is the development of a telemetry device, allowing non-medical staff on board of aircraft to send the data to a specialised diagnostic centre on ground. In this way, a number of unnecessary and very costly emergency landings (a rato of 500,000 EURO on average, as globally accepted figure including potential claims of the other passengers and economic losses) could be avoided.

Two other alternative proposals can be followed to remedy the situation, namely:

- Alternate financing sources
- Public-Private Partnership.

ALTERNATE FINANCING SOURCES.

If we exclude development by private investor financing, we can consider a number of alternatives such as:

- Public subscriptions
- Sponsoring
- Merchandising
- Licensing.

Tierney [38] provides some ideas for a "sponsoring" approach to finance a manned mission to Mars:

- Corporate Sponsoring: Coca-Cola spends 40 million $ just to be called the "Olympic Games Sponsor"
- Marketing tie-ins: Models of the spacecraft, T-shirts, etc. is estimated by experts to be a market of minimum 1 billion $ yearly
- "Name that Peak" approach: the polar explorer Shackleton received 34,000 $ to name a glacier "Beardmore" (after the name of his sponsor)

- TV and Film rights: For reference, the Olympic Games collect some 2 billion $ marketing/ TV income for a 2 week event.

A space related example is Coke and Pepsi, which have been fighting each other for exclusivity rights in space. Coke has spent more than 1,5 million $ on this effort, developing a soda dispenser for use in space, whereas PEPSI has been concentrating on MIR, flying a 1.3 meter module of a Pepsi can outside MIR [39]. Similarly, press articles suggest that "Pizzahut" has paid one million $ to have a large sticker on the Proton rocket which lifted off on 12.7.2000, carrying the ISS Zvezda module.

But also the "space objects" market is rapidly growing, with official auctions and catalogues. The market is presently estimated to be in the order of 20 million $ yearly. As an example we can mention small parts of Moonrocks or even dust collected from returned Apollo suits. A "market" price for such items is in the order of 10 million $/kg.

Extreme alternatives are the creative financing schemes, based upon historical precedents, suggested by Harris [40], a space psychologist, such as:

- A lottery, which was successfully used in England in 1612 to finance the Jamestown settlement and in the 19th Century in America to open the Western Frontier
- Offering of public bonds, as done with the COMSAT offering on the stock exchange (which was oversubscribed)
- Financial incentives to the public, such as tax rebates for space donations.

The underlying aspect of these proposals is evidently to promote voluntary public contribution rather than the obligatory straight taxation system for public space funding.

1.9 PUBLIC-PRIVATE PARTNERSHIP

PRINCIPLES

On several occasions dualism was noted in that:

- The private sector is able to detect early space business opportunities
- The investments and associated risk are too high for the private companies
- Private financing for such high-risk projects is difficult to find
- Technological know-how is available in the public sector
- The public sector is less interested in operating systems.

The importance of public financed research was emphasised again in the results of a U.S. study in 1997 [41]. The study concluded that, although private industry in the U.S. has considerably surpassed government spending on research, still more than 70 % of patents are based upon public research.
A further breakdown of patent sources was detailed as follows (based upon 2,841 patents issued in 1993 and 1994):

- 52.1% based upon academic research
- 11.0% resulting from Government Laboratories
- 10.2% from nonprofit research institutes
- 26.7% only from industry.

Whereas the absolute figures are somewhat surprising, the underlying rationale is less of a surprise: industry focuses its R&D budgets more on the development of (existing) products rather than on fundamental research, highlighting again the important role for research in the public sector. So industry still relies strongly on public funds to perform research and develop key technologies, because they are not able to invest or attract financing to do this on their own.

On the other hand, industry has a broader view on commercial opportunities, and therefore should be involved in an early design phase for such technology programmes. Industry will follow a more commercial Life Cycle Cost (LCC) approach looking at the overall costs associated. This may lead for example to a slightly larger number of less complex satellites, making some "reserves" to compensate the slightly lower reliability (e.g. 90%, instead of 95%, which could represent a satellite at half of the price due to the low mass/low launch cost). Also aspects such as "In Orbit Maintenance" could be brought into the design; for this also, industry needs an early involvement.

In order to merge both public and private interests, more and more Public-Private Partnership (PPP) schemes are envisaged as a mixed solution. The principle is explained by the following rationale:

The initiative of space projects with strong commercial potentialities comes frequently from the industrial and private sectors and is simply based upon the fact that they know the markets and market evolution better. On the other hand, the role of the public space entities remains focused on upstream technological and infrastructure programs. However, development of new space markets sometimes requires heavy front-end investments (for the setting up of the space segment), which requires involvement of the public/governmental authorities. In such cases, the public and the private sectors are called upon to join forces and collaborate in the definition, design and development of a space system with potentially wide-ranging capabilities responding both to the needs of public authorities and to those of private commercial entities. The concept runs in two steps:

- In the first step, the private sector pools its know-how and/or financial resources with the public sector in order to establish a new product, service or technology (development and validation phase)
- In the second step, the venture product, service or technology is mature enough to be marketed to public and/or private clients.

It has to be noted here that we rather should talk about a "PPP-revival" than of a brand new concept. The PPP concept can be found in Dutch literature from around 1986-1988, when it suddenly appeared in the Government objectives of the second Government of Prime Minister Lubbers. Three prerequisites were considered essential for a successful PPP concept [42]:

1. The interest of both parties had to be durable and on a long-term basis.
2. Both parties need to be involved in the earliest possible conceptual phase.
3. The project has a complexity, which discourages private investors.

Otherwise, without these three prerequisites, private investment is considered the more optimal solution for governments.

Also the strategic plan of CNES, in 1996, was designed to revitalise exchange between CNES and other players on the space stage on a partnership basis. It has to be noted here that partnership in this sense is seen in a broader context and covers also areas involving, besides industry, users and other institutional entities (e.g. ESA). One of the CNES examples quoted in [43] is the development of the low-orbit Proteus multi-mission platform, which was a joint CNES-Aerospatiale project. This cooperation could eventually lead to a more important partnership between CNES and industry in the Skybridge satellite communication system (whereby the Proteus platform could be used as a basis for the planned 64-satellite constellation, in a 1500-km MEO).

EXAMPLE OF A PPP APPROACH IN EUROPEAN SPACE ACTIVITIES: GALILEO

A good example of this approach in Europe is the Galileo project, launched jointly in 1999 by ESA and the European Commission (EC) in order to develop a European Global Navigation Satellite System, and to maintain independence of the (U.S. military) GPS. The feasibility of the development is contained in the EC report [44] and as a result a Task Force was set up to come up with an implementation plan for Public Private partnership and submitted to EC Commissioner N. Kinnock on 4th June 1999. The main findings were that Galileo is considered to be an excellent PPP candidate and in line with the guidelines worked out by the European Commission on PPP approaches in general [45], namely:

- That private sector involvement should begin at as early a stage as possible so that they can participate in project designs
- That the public sector should seek, as much as possible, to specify requirements in terms of output (service levels) rather than detailed specifications
- The most effective structure for a PPP normally involves a specially created company vehicle, clearly accountable for project delivery, and with the management autonomy to run an efficient project; and
- That risk shall be allocated according to the capability to control them. This would, for example, mean that the private sector should be responsible for construction cost over-runs, while the public sector would be responsible for cost increases caused by regulatory changes.

As an illustration, we can deduce the following from the report [44]:

Strategy considerations

Galileo will be designed to be an open and global system, fully compatible with GPS, but independent from it, with a role for Russia. They report this consideration upon the existence of an operating system (GPS), which, however, is U.S. Military property and operated by them. Russia, on the other hand, operates the GLONASS system. Therefore, cooperation with Russia could lead to a more rapid implementation (using Russian know-how) and sharing of the GLONASS frequency allocations. Note that, independent of economical problems, Russia has demonstrated maintenance of GLONASS by launching three new satellites at the end of 1998. Further inclusion of, for example, Japan could lead to a worldwide second generation Global Navigation Satellite System, presently called GNSS-2.

System Architecture

As a design driver a three-dimensional performance over land, with an accuracy of better than 10 m horizontally, has been taken. The option of using a core satellite system at MEO (Medium Earth Orbit, between 5,000 and 20,000 km) is presently maintained, consisting of 21 (minimum requirement) or 36 (full independence reached) satellites. A complement of 3 to 9 GEO (Geostationary Orbit) satellites will be needed to coordinate real-time information.

Note that an increased accuracy and reduced response time in the design parameters strongly influence the application range as per table 1.15, adapted from [46].

	< 10 CM	1 M	5 M	10 M	> 50 M
30+ s	Maps Geodesy	Train/Ship positions	Emergency services	Planes (continent)	Planes(over ocean)
10 s	Agriculture	Fluvial navigation	Tracking	Coastal ships	Terminal operations
5 s		Train control	Car navigation	Plane control	Rural transport
< 2 s	Plane collisions	Car theft devices		Maritime collisions	

Table 1.15: Navigation applications in function of accuracy and response delay

The system will be integrated into a cohesive Trans-European positioning and navigation network of ground stations. In a first phase, scheduled for 2003, a system of three Geostationary satellites is envisaged to complement and join the U.S. GPS and Russian GLONASS systems, as shown in fig. 1.7.

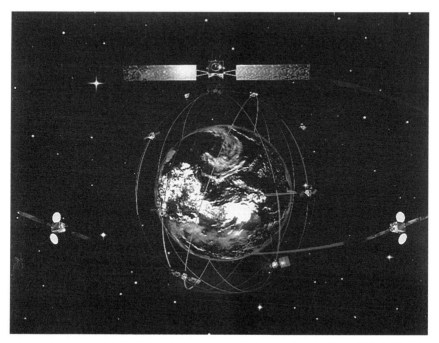

(Photo ESA/J. Huart)

Fig 1.7: Artist impression of Galileo

Cost estimates

The costs for the system depend on the technical solutions chosen, such as the use of the Russian infrastructure or a totally independent development. Another important factor is the choice of a low-level configuration with 21 MEO satellites or a full-scale development with 36 satellites.

Taking these boundary conditions into account the estimated cost figure varies between 2.2 and 2.95 billion EURO, for the development and validation period between 1999-2008. Recurrent (operations) costs, beginning in 2008, are estimated to vary between 140 to 205 MEURO yearly under similar assumptions. A breakdown of the cost is provided in table 1.16.

COST ELEMENT	DEVELOPMENT AND VALIDATION (2000-2005)	IMPLEMENTATION (2005-2008)	TOTAL (2000-2008)
System Engineering	52	90	142
Space Segment MEO	276	592	868
Space Segment GEO	25	163	188
Ground Segment	121	131	252
Operations	55	80	135
Certification	60	189	249
Security	32	31	63
Access Service	100	200	300
Total	721	1476	2198

Table 1.16: Estimated development cost Galileo (in 1999 MEURO)

Financing

A four phase financing scheme is envisaged:

- The first phase (definition/design) to be financed by public funds (EC/ESA)
- The second phase (development/validation) could be based on public funds together with loan guarantees (from EU Member States, European Investment Bank) and minor private funding
- In the third phase (development) funds would be supplied by strategic investors, commercial banks, possibly with public interest rate subsidies
- In the fourth phase private operators are expected to enter the scheme.

Benefits

The various revenue streams underline the interest for both the private as well as the public sector. Whereas commercial applications are evident (e.g. car navigation, fleet management, railways, land survey, commercial and recreational marine) there are a number of macro-economic benefits such as:

- Improvement in response and search time for emergency services
- Enhancement of citizen safety (aircraft crash avoidance systems, anti-theft systems)
- Reduction in traffic congestion and pollution (Intelligent Traffic Control Management Systems).

Financial income will strongly depend on the way in which revenues are levied; one example could be to put a levy on each end-user via smart cards. Assuming for example that by 2010 around 50% of new cars would be equipped with Galileo-based positioning devices, for around 14 million car sales yearly then a 20 EURO levy per car would raise 140 MEURO. Overall assessments estimate the potential market during the operational lifetime at some 80 billion EURO, equally split over equipment sales and value added services.

As an additional macro-economic factor, the positive effect of direct, indirect and induced employment has to be mentioned, covering labour directly associated with the development and the operations of the system proper and the service providers and producers of receiving equipment. An assessment in [44] calculates this employment to be in the order of 240,000 new jobs generated for the whole period 1999-2013.

However, projects of this size are, however, strongly dependent on the macro-economic environment as well, including legislation and political influences. As one example of this interrelation, the American GPS system has a theoretical accuracy of 20-30 meters and until now, the signal was encrypted to degrade the accuracy within 100 meters for civilian use (the so-called "Selective Availability" feature). The decision to abandon this encryption, made in May 2000 [47], is certainly a considerable factor that could influence the decision of a European autonomous system.

It is interesting to note that the previously mentioned "Dual-Use" principle in the U.S. (whereby military based know-how is transferred for commercial purposes), hinders some PPP concepts in the U.S.. Already at a very early phase, the concept has been forwarded (called Public-Private Ventures in the U.S.) for consideration [48] but the involvement of the military projects causes problems with intellectual property rights and with aspects of secrecy.

This observation is formulated in a study on how U.S. Defense research laboratories have problems in adapting to a changing open environment [48]. Although the need is clearly identified to become more market oriented, after the relaxation of "Cold War" imperatives, the fundamental philosophy and culture of such organisations is not facilitating the introduction of marketing concepts. The confidentiality practices of such organisations are only relatively compatible with the required attitudes and customer service practices. The issue of documents and information, requiring the traditional security screening tools, is causing delays and the commercial customer is reluctant to operate in such an environment. The Dual-Use concept has been studied in detail in [49]; in 1996, the following U.S. military R&D budgets were identified:

- 6.2 billion $ on satellite communications
- 8-11 billion $ on Earth Observation
- 12 billion $ on GPS.

It is evident that the reuse of these technologies has given, besides the well-known GPS case, an important boost to space activities, as shown in the article for some examples in table 1.17

MILITARY APPLICATION	COMMERCIAL APPLICATION
Leasat communication satellites	Platform used in Intelsat 6 series.
Milstar communication satellites	Ka-band communications
Modulation technique for communications	Cellular telephones
Deployable antennas for satellites	Digistar broadcasting satellites
NATO communication satellites	Platform for CS-2/3 communication satellites
High-resolution cameras	EarthWatch remote sensing

Table 1.17: Commercial transfer via the Dual-Use effect in U.S.

ENTERPRISING NONPROFITS

In all these considerations, the real rationale behind nonprofit organisations should not be forgotten. Indeed, looking into this evolution, the question arises "Why not go one step further and turn the nonprofit into a for-profit organisation?". Some authors warn against the proverbial 'tail waggling the dog' effect [50] and list the dangers of "Enterprising Nonprofits" as follows:

- New sources of revenue can pull the organisation away from its original mission. If profit becomes a driving force, a number of ESA principles ("Buy European", "Industrial return") could not be maintained anymore
- Commercialisation can undermine the role a nonprofit organisation plays in society. Questions like "What is the profitability of Space Science?", "Who would risk investment in space exploration?" may arise
- Many nonprofit organisations, moreover, do not have the managerial structures to survive in a commercial environment
- Early and applied research are a priori nonprofit activities, because there is no profit possibility. Society would run out of "cutting-edge" products without nonprofit research organisations.

In case the decision is taken to enterprise a nonprofit or government entity, following remedial measures are suggested:

- Create a number of favourable conditions (flexibility, reward systems)
- Identify the business opportunities (which particular strengths can be developed into commercial activities?)
- Assign the venture responsibilities to a person that understands industry and the commercial environment
- Apply for-profit mechanisms (cost control, price determination).

One solution could be to establish a separate entity, preferably even with a separate legal structure.

1.10 CONCLUSION

A market with a yearly turnover of 100 billion $ and employing worldwide more than one million people cannot be ignored. Moreover, if the expected annual growth of 9 % becomes a reality, the Space Market will increasingly grow in importance.

Even if they are not visual at first sight, many day-to-day applications are becoming more and more dependent on this space market, providing a range of products ranging from "handy's" to satellite TV and car navigation systems. This has led to the gradual entry of industrial suppliers in the market, who are basically interested in commercialising the end-products but, in parallel, show more and more interest in end-to-end markets.

The investments and associated (legal) uncertainties are still very considerable. Therefore, it is difficult to find private investment capital to finance such endeavours. Industry is forming larger entities by verticalisation and mergers, which to a certain extent, allows them to carry such important investments on their own risk. With this private investment, new financing methods, such as sponsoring and merchandising, are also appearing on the scene. Public financing is still an important factor because access to space will require a considerable reduction in the costs for upload and important investments in basic technologies are necessary in order to reach such goal.

One interesting form of mixed financing is the Public-Private Partnership. In this PPP scheme, early investments are carried by public funds but involve the private sector in the design and development phase. The operational product, which is the result of this development, is supposed to be closer to the market demand and can be carried further by private investors. In this way, public funds can be made available again to start new space endeavours.

REFERENCES CHAPTER 1

1. OBERTH, H., *Die Rakete zu den Planetenräumen* (Oldenbourg Verlag, Muenchen 1923).
2. RYCROFT, M., *The Cambridge Encyclopedia of Space*, (Cambridge University Press, Cambridge, 1990) p.12.
3. ISBC, *State of the Space Industry 1999*, (Sage Publ., Bethesda, 1999)
4. NAISBITT, J., *Global Paradox* (Avon, New York, 1995) p. 133.
5. BATES, J., Satellite Internet Market Expanding Rapidly, *Space News,* (April 24, 2000) p.4.
6. EUROCONSULT, *The European Space Industry* (Sevig Press, Paris, 1991)
7. EUROSPACE, *European Space Directory 1998*, (Sevig Press, Paris, 1998).
8. SAUL, K., How National Space Activities May Be Integrated Into the Economic Mainstream. In HASKELL, G. and RYCROFT, M., *New Space Markets* (Kluwer, Publ., Dordrecht, 1997), pp. 23-30.
9. HARVEY, B., *The Chinese Space Programme,* (Wiley, Chichester, 1998).
10. EUROSPACE, EFFECTS of the long-term ESA programme on employment, *Space Policy*, Vol.3 No1 (February 1987) pp. 52-64.
11. LIONNET. P., *The European Space Industry in 1998* part 4, www.eurospace.org (March 2000)
12. WOODCOCK, R., On the Economics of Space Utilisation, *Raumfahrtforschung*, 3 (1973), pp. 135 - 146
13. ASHFORD, D., The Commercial Demand for Space Stations, *Journal of the British Interplanetary Society*, Vol. 44, pp. 269 - 274, (1991)
14. ESA: *SE&U Study report* (DLR, Cologne, 1999)
15. BERGER, B., Asian Market Booming for Space Burial Firm, *Space News*, (September 13, 1999), p.4.
16. EUROPEAN COMMUNITY, *Crossroads in Space*. EUR 14010, (EC, 1992).
17. ESA, *Invest in Space* ESA-SP-2000 (ESA Noordwijk, 1999).
18. TROTT, P., *Innovation Management & New Product Development,* (Financial Times Publ., London, 1998).
19. GREENBERG, J. and HERTZFELD, H., *Space Economics*. AIAA Vol. 144 (AIAA, Washington, 1992), pp.323-356.
20. BONNET, R. and MANNO, V., *International Cooperation in Space*. (Harvard University Press, London, 1994).
21. COHENDET, P., Trends in Space Markets, in HOUSTON, A. and RYCROFT, M. *Keys to Space,* (McGraw-Hill, New York, 1999).
22. LOGSDON, J., Space Organizations and International Cooperation, in HOUSTON, A. and RYCROFT, M. *Keys to Space,* (McGraw-Hill, New York, 1999).
23. VON BRAUN, W., *Space Frontier*, (Holt, New York, 1967) p.199.
24. HAYS, J., Three Systems in Search of a Market, *Space News*, (October 11, 1999) p.1.
25. LIONNET, P., The European Space Industry in 1998 part 5, www.eurospace.org (March 2000)
26. ALEXANDER, B., *1996 year in Review*, (ANSER, Arlington, 1997) p.5.
27. KEESLING, G. and KAYNAMA, S., Nonprofit Organizations' Strategic Responses to Changes in Their Funding Environment. *Journal of Nonprofit & Public Sector Marketing*, Vol.4 nr. 1 / 2 (1996), pp. 161-176.
28. SPACE NEWS, Space News Top 50, *Space News,* July 19 (1999), p.8.
29. KAYSER, V., Liability Risk Management for Activities relating to the Launch of Space Objects. *Doctorate Thesis*, (McGill University, Montreal, 15 May 2000).

30. HENNESSEY, J., *Global Marketing strategies*, (Houghton Miflin, 2nd ed., 1992), p. 273.
31. LEVITT, T., The Globalization of Markets, *Harvard Business Review*, May-June 1983 pp.92-102.
32. BOECKSTIEGEL, K.-H., Legal Framework for Privatising Space Activities, *Proc. of the Project 2001 Workshop*, Vienna, (July 1999).
33. SIMONOFF, J., in *The Cambridge Encyclopedia of Space* (Cambridge University Press, Cambridge, 1990) p.339.
34. HALGOUET, DU M., Financing space ventures: innovative approaches, in: *Space Technology and Opportunity* (Online Publ. Pinner, UK), 1985 pp. 115-131.
35. O'DALE, C., Establishing an Infrastructure for Commercial Space, *Journal of the British Interplanetary Society*, vol. 50 (1997) pp. 43-50.
36. KREISEL, J., Space Finance Today and the Need for 'Spacialized' Venture Capital Ignition Space Ventures. *Paper presented at IAA-99*, Ref. IAA-99-IAA.1.4.0.4 (AIAA, 1999).
37. DE SELDING, P., CNES To Finance High-Tec Start-Up Companies, *Space News*, (May 29, 2000) p.4.
38. TIERNEY, J., How to get to Mars (And Make Millions!), *The New York Times Magazine*, (May 26, 1996) pp. 21-25.
39. DUNN, D., The Final Frontier: Cola War in Space, *Moscow Tribune* (May 22, 1996).
40. HARRIS,P., *Living and Working in Space,* (Ellis Horwood, New York, 1992), p.292
41. Broad, W. Study finds Public Science is Pillar of Industry, *New York Times*, (May 13, 1997).
42. VERGEER, A., Financiele Aspecten van Public Private Partnership. (in Dutch) *Financieel Overheidsmanagement,* (4, 1988) pp.4-10.
43. CLERC, P., Partnership between Cnes and Industry: A New Market Oriented Approach. In Haskell, G. and Rycroft, M., *New Space Markets* (Kluwer, Publ., Dordrecht, 1997), pp. 13-22.
44. EUROPEAN COMMISSION, Galileo. Involving Europe in a New Generation of Satellite Navigation Services. *Report COM(1999)54*, (Brussels, 10 February, 1999)
45. EUROPEAN COMMISSION, Public-Private Partnership in Trans European Networks. *Report COM (1997) 453*, (Brussels, 10 September 1997).
46. BORIES, B., Galileo, La Navigation Européenne par Satellites, *La Revue des Télécommunications d'Alcatel* (4/1999) pp. 286 – 291.
47. ROBBINS, C., U.S. Loosens Rules on GPS, *Wall Street Journal Europe,* (May 2, 2000), p.27.
48. BUYS, K. and HANNA, J., The Compatibility of the Marketing Concept Within the "Classified" Customer Service Culture of Department of Defense National Testing Laboratories, *Journal of Nonprofit & Public Sector Marketing,* Vol.5 (4), 1997, pp. 49-68.
49. GIGET, M., CHENARD, S. & LE PROUX DE LA RIVIERE, E., Impact of the dual use concept along the value chain of major space applications, *Acta Astronautica,* 38 (1996) pp. 587-603.
50. DEES, G., Enterprising Nonprofits, *Harvard Business Review,* (January 1998), pp.55-67.

Chapter 2

USE OF A SPACE NONPROFIT MARKETING MIX

2.1 INTRODUCTION

After an initial period of discussion if "classical" marketing tools could be implemented in the nonprofit sector, this topic by now should have been positively concluded in general terms. Most profit organisations, faced with competition from similar or commercial entities, have established professional marketing departments and are using the marketing techniques, which have matured in the commercial sector over many years. The simple approach of a Marketing Mix consisting of the so-called 4 Ps (Product, Price, Physical Distribution and Promotion) is now widely accepted.

In a second step of refinement, however, it becomes clear that the tools of the classical Marketing Mix cannot be used without tuning to the specific nonprofit environment. The new concept of extended Ps, compared to the original 4 Ps, offers new perspectives and possibilities for this. The variety of nonprofit Organisations does not allow for selecting one single approach because each organisation has to tailor its marketing mix towards its target groups. Therefore, the approach followed traces a number of steps from the past on how the concepts have evolved, namely:

- The historical evolution of classical marketing
- The transition from classical to nonprofit marketing
- A definition of the nonprofit sector and classification
- Analysis of the differences between profit and nonprofit market
- Potential criticisms of nonprofit marketing
- The broadened concept of marketing tools
- and, finally, the concept of an appropriate Marketing Mix for Space Marketing.

The reason why this chapter is introduced here is evident; marketing is used, in general, when the pressure of competition is felt and resources to be distributed get scarcer. During the "Space Race" era these prerequisites did not exist and Space Organisations still have the tendency to continue operating under these conditions. On the other hand, specifically in the scientific world, there is still some resistance against the use of Marketing techniques because they are automatically related to consumer products.

For both reasons, it was felt of importance to look back into the historical evolution of marketing in the public sector, as such an evolution is typical for each sector; only the timing has been different in the past.

This chapter emphasises nonprofit marketing because in this case most argument is still needed. Marketing instruments in the space business environment have been practised for since a long time and are fully covered by industry.

2.2　MARKETING IN ITS HISTORICAL CONTEXT

Early marketing roots can be found already in the works of Adam Smith, who remarked, as long ago as 1776, that [1]:

Consumption is the sole end and purpose of all production; and the interest of the producer ought to be attended to, only as far as it may be necessary for promoting that of the consumer. The maxim is so perfectly self-evident that it would be absurd to attempt to prove it. But in the mercantile system, the interest of the consumer is almost constantly sacrificed to that of the producer; and it seems to consider production, not consumption, as the ultimate end and object of all industry and commerce.

Nevertheless, till the 1920's, industrial Organisations were purely **production oriented.** The philosophy was to "sell what you can produce"; the customer's wishes were not considered (a typical example being the Ford Model-T automobiles, for which, Henry Ford claimed, they could be delivered in any colour, as long it was … black).

During the recession of the 1920-1930's, there was a sudden reduction in demand. The philosophy turned in the direction of "produce what you can sell" and Organisations became **sales oriented**. After the Second World War, a number of elements reinforced this process:

- There was an overcapacity of production means. Airplanes, cars and even the (Liberty) ships had been in mass production for years and a lot of capacity was now standing idle
- Women had been involved in production, so that, after the War was over, there was a considerable surplus of skilled labour
- Frontiers opened; in the early 1950's European companies were confronted with U.S. and new European companies entering the market, whilst Japanese products were entering the world markets.

The consumers were offered a wider choice of products and producers from whom to purchase. Consumers were exercising power on the producer and elected to purchase only from those whom they felt would adequately service their needs. With this transition from a "seller's market" to "buyer's market", producers were forced to adapt their approaches and they became more and more **customer oriented**. Companies became aware that the sales function was not the only one, they also had to "sell themselves".

The idea of a "Marketing Mix" was born. McCarthy [2] introduced this in the 1960's with the simple – and now famous – 4 Ps, meaning that each Marketing Approach had to be based on four pillars: Product, Price, Place and Promotion. At this stage, we first have to introduce more exact definitions. Different writers have defined marketing in various ways. As quoted in [3], in 1960, the "American Marketing Association" (A.M.A.) defined, in the "Report of the Definitions Committee", marketing as:

The performance of business activities that direct the flow of goods and services from producer to customer.

In probably the most popular book on Marketing, Ph. Kotler [4] prefers the definition

Marketing is a social and managerial process by which individuals and groups obtain what they need and want through creating and exchanging products and value with others.

In line with this, the Marketing Mix is defined as

The set of marketing tools that the firm uses to pursue its marketing objectives in the target market.

It has to be noted here that Kotler refers to a number of other definitions. Already in 1975, Crosier [5] reviewed over fifty different definitions of "marketing", which he classified into three major groups:

- Those that regard marketing as a "process"
- Those that see it as a "concept or philosophy of business"
- Those that regard it as an "orientation".

We should mention in this context that Kotler recently often also uses the more direct definition [6]:

Marketing is the science and art of finding, keeping, and growing potential customers.

In the framework of the historical developments, one cannot overlook an early article of Levitt in 1960 [7], which already introduced the term "Marketing Myopia". He stressed the concept of serving and satisfying human needs, with the ultimate objective to focus on the customer's needs. The short-term objective is to sell existing products to people; long-term survival lies in the creation of products that people need. Sellers who overconcentrate on the product instead of the customer's need are considered to suffer from such marketing myopia.

There is one interesting aspect to this, namely that a clear tendency can be noted that broadens the scope of marketing, as reflected in the equally broader definitions. In particular, one later definition makes the first bridge to other sectors, namely the one from Kotler and Fox [8]:

Marketing is the analysis, planning, implementation and control of carefully formulated programs designed to bring about voluntary exchanges of values with target markets for the purpose of achieving organizational objectives. It relies heavily on designing the organization's offerings in terms of the target market's needs and desires and to using effective pricing, communication and distribution to inform, motivate and service the markets.

In addition to product oriented marketing, this definition broadens the scope to two fields, which developed in parallel, namely Service Marketing and Nonprofit Marketing.

Kotler defines a service as [9], p.376:

A service is any activity or benefit that one party can offer to another that is essentially intangible and does not result in the ownership of anything. Its production may or may not be tied to a physical product.

Based upon this distinction with the traditional Product Marketing, Service Marketing has developed over the last few years, but mainly focuses on the profit making part of service industry. Indeed, Zeithaml and Bitner [10], p.5 describe the field of service marketing as:

All economic activities whose output is not a physical product or construction, is generally consumed at the time it is produced, and provides added value (such as convenience, amusement, timeliness, comfort and health) that are essentially intangible concerns of its first purchaser.

Earlier, Levitt [11] had already introduced the intangibility element. He defended the position that each product has an intangible side, also the physical products; the main difference between the physical products and services being the degree of intangibility.

THE ORIGINS OF NONPROFIT MARKETING

An important milestone in nonprofit marketing was the article of Kotler and Levy from 1969 on "Broadening the Concept of Marketing", [12], with the often quoted extract:

It is the author's contention that marketing is a pervasive societal activity that goes considerably beyond the selling of toothpaste, soap, and steel. Political contests remind us that candidates are marketed as well as soap; student recruitment by colleges reminds us that higher education is marketed; and fund raising reminds us that "causes" are marketed. Yet the student of marketing typically ignores these areas of marketing. Or they are treated cursorily as public relations or publicity activities. No attempt is made to incorporate these phenomena in the body proper of marketing thought and theory. No attempt is made to redefine the meaning of product development, pricing, distribution, and communication in these newer contexts to see if they have a useful meaning. No attempt is made to examine whether the principles of "good" marketing in traditional product areas are transferable to the marketing of services, persons, and ideas.

The article was clearly intended to be provocative, as can be illustrated by the following example quoted in the referred article:

The junta of Greek colonels who seized power in Greece in 1967 found the international publicity surrounding their cause to be extremely unfavorable and potentially disruptive of international recognition. They hired a major New York public relations firm and soon full-page newspaper ads appeared carrying the headline "Greece Was Saved From Communism", detailing in small print why the takeover was necessary for the stability of Greece and the world.

The provocation worked, and one can follow a vivid debate on the applicability of "classical" marketing concepts to the nonprofit sector. Many academic marketers attacked the idea, stating that marketing only made sense in profit-oriented enterprises. Nevertheless, in the 1970's this gave rise to early applications of marketing techniques to fund raising, health services, and other aspects of social marketing. The July 1971 issue of the Journal of Marketing was even dedicated to such applications.

As another important milestone in the awareness process, in 1973 Shapiro [13] provided an analysis of the major marketing tasks for the nonprofit manager, which he describes as

- resource attraction
- resource allocation
- persuasion,

elaborating how the 4Ps of the classical marketing mix can be used in the performance of these tasks. In 1975, Kotler published the first full text on "Marketing for Nonprofit Organizations", which has been regularly reworked and reprinted, remaining the standard textbook on Nonprofit Marketing [9].

Hunt proposed a conceptual model with separate categories for the profit and nonprofit sector, concluding that [14]:

My own belief is that the similarities between marketing in the profit sector versus the nonprofit sector greatly outweighs the differences.

The initial debate on the differences seemed to prevail over the debate on the similarities, until finally, in the early 1980's emphasis was placed on how to develop effective marketing for public and nonprofit Organisations.

It was again Kotler in another – now classic – article [15] who defended the concept against continuing resistance. He described and countered a number of problems, such as:

- Many groups within the nonprofit organisation saw marketing as a threat to their autonomy and powers. He correctly pointed out that this is a "resistance to change", which will be overtaken by the facts
- Overemphasis that was initially placed on promotion and advertising
- Some observers saw a risk that only major institutions would be able to spend enough resources to "market" properly, fearing that smaller ones who could not afford marketing would disappear. For Kotler, specialisation in various market segments is the logical solution to this.

In the meantime, the advantages of Marketing in nonprofit Organisations have been generally accepted. Recently, Sargeant [16] summarised these advantages as:

1. Marketing can improve the levels of customer satisfaction attained
2. Marketing can also assist in the attraction of resources to a nonprofit organisation
3. The adoption of a professional approach to marketing may help an organisation to define its distinctive competencies, by defining what an organisation can offer society that others cannot
4. Professional approaches to marketing also offers organisations a framework within which to effectively work, minimising the waste of valuable resources.

2.3 DEFINITION OF THE NONPROFIT SECTOR

Defining the Nonprofit Sector is not easy, because there is a very high similarity with commercial services, specifically in the provision of services (e.g. postal service, telephone services, and public transport). In fact, the borderlines between both seem to become more vague recently, when compared to activities in preceding periods. A second aspect is the heterogeneity of the activities involved. Kotler [9] p.12 refers to following parameters:

- **Main source of funding**: Profit as income for the Profit Organizations, Government funding and donations on the other hand for the Nonprofit Organisations
- **Types of products**: Tangible products versus Services and behaviour changes
- **Extent of political control**: From "tight political control" to "largely independent of political control".

A recent definition is introduced by Sargeant [16] p.4, being:

> *A nonprofit organization may be defined as one that exists to provide for the general betterment of society, through the marshalling of appropriate resources and/or the provision of physical goods and services. Such Organisations do not exist to provide for personal profit or gain and do not, as a result, distribute profits or surpluses to shareholders or members.*

Note: we should mention here that the term "nonprofit" is not generally accepted (as this, literally, can also be applied to organisations bearing a net loss). Purists, therefore, prefer the use of the term "not-for-profit organisations". We will, however, adhere to the more generally used term "nonprofit". In a literature search, the problem is even made worse by the spelling as "non-profit", which is used in American literature.

Various classification systems have tried to identify and describe these organisations in more detail as a function of their nature. One of the commonly used distinctions is the one made by Hansmann [17], in accordance with two sets of characteristics:

1. The organisation is **donative or commercial**; that is, revenues are primarily either secured by donations or from charges/fees to users.
2. The organisation is **mutual or entrepreneurial**; whereby the mutual organisations are managed by the users themselves and the entrepreneurial ones by professional and paid managers.

Distinctions are never unambiguous; also profit organisations must render services to their clients which are not directly profit making (advice, guaranteed repairs), whereas nonprofit organisations also receive money from proper sales (e.g. the sale of pins and T-shirts, third party services).

Furthermore, the nature of the organisation does not always allow an a priori classification: some government corporations follow purely profit-economical philosophies (port authorities, state airlines) and, on the other hand some hospitals and schools are organised as profit oriented organisations. Since 1993, the Dutch Post Offices (a purely state-owned organisation), have sold commercially products as sweets, gifts, stationary and gadgets in their offices in order to share the operations cost of these offices (operations had fallen rapidly due to the influence of credit cards and tele-banking). Thereby we note a very typical example off market-diversification in the nonprofit sector. However, most of the principles that apply to nonprofit organisations apply to such "borderline" organisations as well.

Therefore, Lovelock and Weinberg [18] p. 31 have therefore proposed a more flexible presentation (figure 2.1), based upon Hansmann's distinction and put the dimensions as axes, which allows one to position the respective organisations within these axes including nuances.

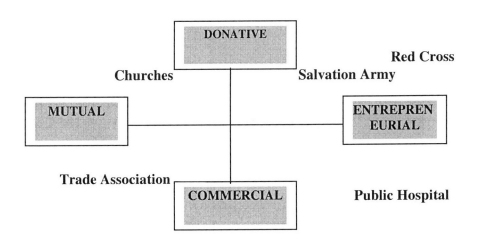

Fig. 2.1: Types of nonprofit organisations (with examples)

An international classification is given in the ICNPO (International Classification of Nonprofit Organizations) system. The first layers of this classification are given in the "subgroups" of table 2.1.

Group	Subgroups
1. Culture and Recreation	1 100 Culture 1 200 Recreation 1 300 Service Clubs
2. Education and Research	2 100 Primary and Secondary Education 2 200 Higher Education 2 300 Other Education 2 400 Research
3. Health	3 100 Hospitals and Rehabilitation 3 200 Nursing Homes 3 300 Mental Health 3 400 Other Health Services
4. Social Services	4 100 Social Services 4 200 Emergency and Relief 4 300 Income Support and Maintenance
5. Environment	5 100 Environment 5 200 Animals
6. Development and Housing	6 100 Community/ Social Development 6 200 Housing 6 300 Employment/Training
7. Law Advocacy and Politics	7 100 Civil/Advocacy Organizations 7 200 Legal Services 7 300 Political Organizations
8. Philanthropic/Volunteerism	8 100 Philanthropic/Volunteerism
9. International	9 100 International Activities
10. Religion	10 100 Religious Congregations
11. Professional Organizations	11 100 Professional organizations and Unions
12. Not Elsewhere Classified	12 100 N.E.C

Table 2.1: ICNPO classification of nonprofit Organisations

Over the years, there has been a change in the nature of nonprofit Organisations. Kotler [9] described the evolution in the following way:

The earliest stage of nonprofit organisations conforms to a voluntary/civic model. From the earliest times until the end of the 19th century, services were rendered by groups of citizens (examples are volunteer fire departments, building of community houses). Most of these services were inspired by religious or personal philosophies based on sharing, and evolved from a low level of economic welfare.

After the (first) industrial revolution, wealth came into the hands of a few families. Due to a sense of social sympathy (or simply out of guilt?) this resulted in an era of philanthropic patronage, specifically expressed in a number of foundations (Rockefeller, Carnegy, and Nobel).

Following the periods of depression and world wars, the nonprofit model turned to a model based on rights and entitlements, whereby social groups claimed to have a right to support from public funds. This model has recently evolved onto the competitive/market stage: nonprofit organisations realised that they could not fully depend upon:
- the individual willingness to share,
- the generosity of the wealthy,
- the support of governments,

and they tried to achieve more independence. This resulted in more sound management principles, professionalism in staff and ... marketing. This last change is certainly being influenced by a number of new orientations:

1. The government-spending pattern is changing; major cutbacks are introduced.
2. The philanthropic character is decreasing. Although the absolute figures for donations are, due to their nature, difficult to assess, charitable giving, as a percentage of disposable income (pre-tax earnings) dropped from 1.99 to 1.84 % between 1970 and 1980 in the U.S.
3. The economic structure is changing: private organisations "attack" traditional nonprofit sectors such as health care and public transportation.

Extensive research on the macroeconomic importance of nonprofit organisations was done in 1995 by the John Hopkins University covering 22 countries [19] and results for some countries are represented in table 2.2.

COUNTRY	PEOPLE EMPLOYED (X)	PUBLIC FUNDING	INCOME FROM SALES	OTHER INCOME
Netherlands	12.4 %	60%	36%	4%
Belgium	10.5 %	77%	18%	5%
U.S.	7.8 %	31%	57%	12%
U.K.	6.2 %	47%	45%	8%
France	4.9 %	58%	35%	7%
Germany	4.5 %	64%	32%	4%
Japan	3.5 %	34%	62%	4%

(X) = As percentage of total employment, excluding agriculture

Table 2.2: Economic key indicators of the nonprofit sector

In the main industrial countries (U.S., U.K., F, I, D, J) table 2.2 shows that the nonprofit sector contributes some 5% of the GNP on average and employs 11.8 million workers. This last figure is of considerable importance: even if the turnover in financial terms is relatively low, we should not underestimate the effect of Nonprofit Organisations on the labour market. Even more so in emerging countries where for example one single organisation in Sri Lanka has 9000 paid fieldworkers and 41,000 local fieldworkers.

2.4 DIFFERENCES BETWEEN THE PROFIT AND NONPROFIT SECTOR

Essential for the choice of a Marketing Strategy and the right Marketing Mix is a good understanding of the differences between both sectors. Anthony and Herzlinger summarises these as follows [20]:

(a) Lack of profit measure for allocation of resources:

The amount of profit can give an objective measurement of productivity. Without this profit measure, outputs cannot be measured in quantitative terms. This, in turn, makes control of productivity extremely difficult. The consequences of this are more important than one might at first sight expect and result in:

- Lack of single decision criterion: there is often an ambiguous objective that can be used in analysing proposed alternative courses of action
- Difficulty of relating costs and benefits. There is no way of estimating the benefits of a given increment in spending
- Difficulty of measuring performance. Performance with respect to important goals is often difficult to measure
- Comparison of units. In case of budgetary restrictions, how are priorities given?

(b) Tendency to become service organisations:

Goods can be stored, services not. As a consequence, (governmental) organisations tend to have excessively high staff complements, in order to be able to respond to - infrequent - peak loads.

(c) Constraints on goals and strategies:

Most services to be provided are strictly directed by an outside authority rather than by inside management. Most service organisations must confirm to the wishes of fund providers which are not always equal to the wishes of the customers.

(d) Source of financial support:

Contrary to industry, income is independent of the outcome of the services rendered. The relation may even become negative, for instance at periods of budget restrictions the need for social and medical services may increase.

(e) Dominance of professionals:

Specialists (scientists, researchers, military commanders,) often have motivations that are completely inconsistent with good resource allocation. The provision of additional management training courses have become partially remedial actions in recent years.

(f) Difference in governance:

In Industry, the ultimate authority and control is well known, this is not the case for nonprofit organisations. External influences, from various sources, are more likely to result in confusion than in clear guidelines.

(g) Differences in top management

President Truman's desk bore a sign "The Buck Stops Here"; meaning that he could not delegate decisions any further. In most nonprofit organisations there is a multi-headed top management with sometimes ... bucks heading off in all directions.

(h) Importance of political influences:

In general, the prime objective of elected officials is to ... be re-elected. Frequently, this leads to conflict with the interests of the organisations such officials steer. Another important factor is that top management tends to change as a result of changes in administration or shifts in the political climate. This may result in short-term objectives and programmes that produce visible results quickly, rather than longer range programmes.

(i) Tradition of inadequate management controls:

Well-established accounting methods, even those with proven results, have not been introduced in governmental organisations. There is a tremendous resistance to change, which is very hard to overcome.

In this context, it is useful to elaborate on the term "bureaucracy". This term was first introduced by Weber [21], as a classification of a sociological organisation-style: the executive derives his authority by law or by the election of an authorised body, rather than by tradition or charisma (the two other types of organisations in Weber's categorisation). In the bureaucratic organisation, complex problems are solved by segmenting them into a series of simpler ones and delegating authority for solving each of the segments to specialised subunits of experts. Bureaucracy avoids the danger of subjectivity by replacing the subjective judgement of individuals by a set of strict rules, values and attitudes, which are approved and controlled by the expert's superior. Under this definition, some form of bureaucracy is a necessary condition in most large and complex organisations. This scientific definition of bureaucracy should not be confused with the pejorative one used in daily language, being an organisational structure operating by overcomplicated rules and routines, resulting in delays and "buck-passing" without respect for the client's needs.

In predicting paramount changes in society, under the influence of knowledge technologies, even Peter Drucker acknowledged the continuation of bureaucracy in governmental activity [22] when stating,

> *Government is also properly conscious of the fact that it administers public funds and must account for every penny. It has no chance but to be "bureaucratic", in the common sense of the word.*
> *(…) This means, however, that government "bureaucracy" cannot be eliminated. Any government that is not a "government of forms" degenerates rapidly into a mutual looting society.*

An alternative description of the differences between nonprofit and "for-profit organisations" can be found in Lovelock and Weinberg [23]. They reduce such differences to a number of significant ones, which, in their opinion, are important from a Marketing point of view. Their description can be summarised as follows:

CHAPTER 2 : SPACE MARKETING MIX

(i) Multiple Constituencies:

The client, who receives services, and the donor, who provides funds, are often unrelated and from considerably different groups. This results in two often very different markets and market strategies. What attracts the donor and may improve funding can at the same time shock the "client" (for example HIV/AIDS campaigns).

(ii) Non-financial Objectives:

A balance needs to be maintained between financial profit and social profit. Some authors criticise overemphasis of the Price element in a nonprofit marketing plan because success or failure cannot be measured in strictly financial terms only. This argument has led to the creation of other tools, over and above the classical 4Ps.

(iii) Services and Social Behaviours Rather than Physical Goods:

The intangibility of services makes them more difficult to evaluate; this area is, however, improving and becoming well explored under Service marketing. Social-behaviour programmes face another dimension, in that they are even often perceived as controversial as they try to change lifestyle patterns.

(iv) Public Scrutiny/Non-market Pressure:

Most public services are subject to very close public scrutiny, particularly if the main source of income is public money. With our increasing access to information, such scrutiny is more and more emphasised. Moreover often they are not dictated by the "discipline of the marketplace" and subject to political influences, even up to the point of being used as electoral instruments (e.g. public transport availability, choice of trajectories for motorways/trains...).

(v) Tension between Mission and Client Satisfaction:

There are long-term or philosophical considerations which will not always be appreciated by the client (examples: reluctance to use antibiotics in hospitals, new art forms which are not – yet – accepted by sponsors and the general public). The "customer is always right" principle cannot always be followed, as it may even be dangerous on the long-term for the client himself. Such tensions, which require good communication to be remedied, can become very sharp in resource-scarce times.

All these differences explain the difficulties to simply transpose profit making marketing principles in the nonprofit sector. They also explain the risk of having criticism directed against marketing in the nonprofit sector.

2.5 CRITICISM AGAINST MARKETING IN THE NONPROFIT SECTOR

Certainly in Europe, there is still considerable hesitation to employ full-scale marketing techniques in the nonprofit sector. It is important to know the major concerns raised generally in order to adapt a possible counter strategy. Kotler [9] summarises and illustrates these criticisms as follows:

→ **Criticism1: Marketing is a waste of public money**

An example of such public criticism occurred when the American Army spent 10.7 million $ in 1971 on advertising in order to increase enlistment. Such a public reaction is a major problem requiring a rational marketing strategy. Specifically when taxpayers' money is involved, the expenditure for marketing purposes has to be closely monitored. The situation is aggravated in those cases where profit-making competition is appearing, a typical case being some space sectors at present. No organisation should add costs that do not increase results or give an adequate return. Nonprofit Organisations owe their customers at any time an explanation of what they are seeking to achieve through their marketing expenditure. If a Space Agency distributes documentation to schools it will be accepted as educationally valid. If the same material would be distributed via paid TV-commercials the reaction might not be so positive.

→ **Criticism2: Marketing activities are seen as intrusive**

Certainly in a number of "welfare" sectors, such as the health and social ones, marketing research often deals with very personal questions. A number of consequences result from this observation:

- People often do not respond openly, which leads to wrong conclusions
- People are afraid that their answers could be used by governments against individual citizens or for mass-propaganda

- Only certain categories of people will react, other will refuse to respond. Therefore the value of resulting statistics may be unequally influenced.

There is of course a vicious circle in this respect. Market research is primarily carried out to learn exactly what the wishes and attitude of the people are so that eventually the organisation can provide better services.

→ **Criticism3: Marketing is manipulative**

As an example, some aggressive campaigns against AIDS were considered strongly discriminating by certain groups of the population (e.g. homosexuals feared that this could result in even bigger discrimination). Administrators of nonprofit Organisations should be sensitive to the possible charge of manipulation when they implement a marketing programme. One should be aware that there could "back-fire" on the organisation if a marketing campaign ignores this aspect, putting such an organisation on the "black list" for many years.

These criticisms led to a number of restrictions when applying Marketing in the nonprofit sector:

- Budgets for advertising and market research have to be severely scrutinised
- Major use of donated advertising, not paid advertising
- Discourage research into areas of high sensitivity
- Avoid "aggressive" marketing techniques, such as use of "fear" and "hard selling" techniques.
- Establish internal "codes of ethics" and discipline (as applied in medicine and accounting).

Andreasen [24] p.30 emphasises this as follows:

> *(Non-profit) marketers bear a special obligation to behave in an ethical fashion because they are purporting to act in society's interests and not – unlike commercial marketers – in their own. They take as their challenge influencing behaviour in ways that are good for the individual and for the society. This role, I believe, requires that they pay extraordinarily close attention to the ethics of the goals they choose and the means they choose to get there.*

On the goals, emphasis is put on the subjective appraisal of the marketing designer. Therefore, Andreasen strongly advises that decisions with potentially controversial reactions are first "vetted" by some sort of societal representative collective or panel, in order to at least increase the probability that the project outcome will not be counter to popular norms and values. The source of most criticism, however, often lies with the simple observation that two related misconceptions often appear in nonprofit organisations:
- Either, the customer is unknown or wrongly identified
- Or, his wishes are simply ignored.

This reflects a recent evolution that all nonprofit Organisations have to take into account. In the end the customer will determine what he wants and will react with the means, which are at his disposal. The nonprofit sector has been less transparent in the past, but with an orientation to an open, information oriented, society, the general public, supported by the press, will be able to evaluate "cost/benefit" ratios even on government spending.

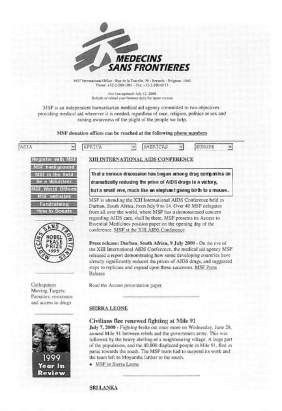

Fig. 2.2: Example of an informative, neutral website of a nonprofit organisation

The general public has too often in the past been confronted with cost and time overruns. They are now becoming more and more vigilant and react themselves. As an example, we can illustrate this with some overruns from big projects in Germany (1 MDM = 1 million Deutsch mark = 500,000 $) :

- In 1927, the building of the Nuerburgring cost 14 MDM, against the 2,5 MDM estimated. People were certainly not happy with this...but no major protest has been recorded
- In 1972, the Olympia stadium, which was estimated in 1967 to cost 500 MDM, finally turned out to cost 2000 MDM. This time reaction was much stronger, there were inquiry commissions on how the money was spent and some officials were even sued
- In 1982, the influence of the public was so strong that the building of the fast breeder reactor SNR-300, in Kalkar, was stopped at an estimated Cost to Completion of 6500 MDM, against the 1700 MDM estimated in 1972.

These examples, all with the same order of overruns, illustrate a considerable change: the taxpayer, assisted by an improved information flow, gradually evolved from an a posteriori critic towards an a priori controller. Specifically in the case of a nonprofit organisation dealing with taxpayers' money, it is essential to identify the client and to respect his needs. In this context we have to raise a word of caution on the interpretation of statistical data and figures. To quote the well-known management expert Peter Drucker [25]:

> ***To obtain its budget, the nonprofit organisation needs the approval, or at least the acquiescence, of practically everybody who remotely could be considered a "constituent". Where a market share of 22 percent might be perfectly satisfactory to a business, a "rejection" by 78 percent of its constituents would be fatal to a budget-based institution***

The importance of the budget cannot be underestimated in this context, because it is the basis for each nonprofit organisation. If a company has a good product, it can obtain a small portion of the market and make a good living from it. However, if this is financed with public money, there could be a majority of people against this same type of business. For example Christo's wrapping of the Reichstag in Berlin in 1996 did not attract major criticism, because it was self-financed without public money.

2.6 THE CHOICE OF A MARKETING MIX IN THE SPACE SECTOR

BROADENING THE MARKETING MIX CONCEPT

The Marketing Mix concept was introduced by Borden [26], as early as 1953, when he compared the marketing process with baking, whereby a number of ingredients have first to be assembled and then "mixed" in appropriate proportions. He developed his idea in the form of a checklist approach.

Other approaches followed and eventually, the 4Ps approach of McCarthy [2] became the most accepted one. In his opinion, the marketing mix can be presented as a vector $(P1, P2, P3, P4)^t$, whereby:

P1 = Product
P2 = Price
P3 = Physical Distribution (or Place)
P4 = Promotion

These four variables are the major tools which are available to each organisation in order to obtain a better place in the "market". The simple distinction in four Ps is evidently an oversimplification but it has the undeniable advantage that it can be easily memorised and systematically illustrated. In a first approach, each of the Ps has been further defined, specifically based upon the proposals made by Lipson and Darling [27]. In this approach each P on its own is a collection of a number of tools. Very generically one can further split the main vector into four subvectors as follows:

Product, P1 = $(P11, P12, P13, P14)^t$, with

P11 = Features and options
P12 = Service and warranties
P13 = Brand name
P14 = Packaging

Price, P2 = $(P21, P22, P23, P24)^t$, with

P21 = Basic or List Price
P22 = Discounts and Allowances

P23 = Credit Terms
P24 = Transport Prices

Place, P3 = (P31, P32, P33, P34)t, with

P31 = Distribution channels
P32 = Location
P33 = Inventory control
P34 = Shipping

Promotion, P4 = (P41, P42, P43, P44)t, with

P41 = Public Relations
P42 = Sales Promotion
P43 = Personal sellings
P44 = Publicity

As an overall result, we have now 16 subinstruments, which the marketing "manager" or the administrator can use, tailored to the specificities of their respective markets. Doyle [28] rephrases this in a more commercial definition as follows:

- Product (consisting of quality, features, name, packaging, services, guarantees)
- Price (consisting of list price, discounts, allowances, credit)
- Promotion (consisting of advertising, personal selling, sales promotion and public relations)
- Place (consisting of distributors, retailers, locations, inventory, transport).

It is evident from the commercial descriptions that these definitions were strongly inspired by the marketing of physical products, or even more so by goods.

Another problem occurred when trying to implement these tools in such sectors as Service Marketing and Social Marketing. The fact that previous research strongly emphasised physical goods became apparent. Managers from profit sectors were hired in the nonprofit sector and tried to apply the principles and techniques they were familiar with. The dangers of this step, which often lead to "errors of the third degree" (i.e. right approaches for the wrong problems) are very well illustrated by Bloom [29]:

> *The relationship between social marketing and conventional commercial marketing may be somewhat like the relationship between football and rugby. The two marketing "games" have much in common and require similar training, but each has its own set of rules, constraints, and required skills. The good player of one 'game' may not be necessarily be a good player of the other...*

As more and more authors in the social and services sectors, as well as in the nonprofit sector, tried to make their requirements suit the 4Ps, a new "school" developed. Essentially, various authors tried to extend the original 4Ps with other elements, which were assumed to be more appropriate for social and services marketing approaches.

Booms and Bitner [30] argued that three extra Ps should be added to the list, namely

P5 = People
P6 = Physical Evidence
P7 = Process

People refers to the consideration of employment, training and motivation of individual members of service staff. It also refers to the personnel, their discretion, appearance, interpersonal behaviour and attitudes.

Physical Evidence refers to environmental factors such as furnishings, colour, layout, and noise level in order to maximise visitor interaction.

Process refers to the policies, procedures, customer involvement, customer direction and flow of activities.

Zeithaml and Bitner [10] pp. 26-27 refine the definitions as:

> *People: All human actors who play a part in service delivery and thus influence the buyer's perceptions: namely, the firms personnel, the customer, and other customers in the service environment*
>
> *Physical Evidence: The environment in which the service is delivered and where the firm and customer interact, and any tangible components that facilitate performance or communication of the service.*

> ***Process:*** *The actual procedures, mechanisms, and flow of activities by which the service is delivered – the service delivery and operating systems.*

Bruce [31] strongly feels the need to introduce another tool, namely,

P8 = Philosophy

Bruce introduced the Philosophy aspect into the Marketing Mix in 1994, in his book on "Charity Marketing".

He defends his choice as follows [31] p.84,

> *Voluntary organisations should be explicit about their values and philosophy and ensure that these are integrated into the marketing mix. Failure to do so will lead to tensions and contradictions between different physical products, services and ideas.*

By philosophy in the Marketing Mix he understands,

> *The explicit recognition of the value-laden approach to be taken or encouraged in the product, be it a physical good, a service or an idea aimed to be beneficiary, supporter, stakeholder or regulator market.*

Evidently, philosophy will be peculiar for each different organisation. Bruce even feels that the philosophy should be based on the needs and wants of the beneficiary customers. As an illustration: do handicapped children support being pictured on posters? Certainly this is very appealing and will move people, even increase donations, but is the real customer served by such creation of a "pity-environment"? Bruce even pleads, for the charity organisations, to put first the philosophical target, before the price and product ones, in order to avoid e.g. too commercial or ethical approaches.

We notice a similarity with space activities. A number of activities cannot, and should not, be justified with (pseudo) spin-offs. If a scientific satellite is used to participate in the exploration towards the discovery of the "origin of life", this should be on the first page and throughout the whole "marketing" concept. Too many times in the past people have been "sold" space projects by artificial and alleged benefits, which afterwards cannot be substantiated.

It should be emphasised here that this aspect is not only valid for services. The success of e.g. solar energy heating systems has an additional dimension compared to the purely rationale economic one. Embley [32] has published a book with the indicative title "Doing Well While Doing Good", describing 50 examples of companies which have developed their business platform on the basis of "doing good". One good example is "The Body Shop", which clearly abstains from testing its products on animals as well as pursuing a philosophy of obtaining products only from non-exploitative sources.

THE APPROPRIATE CHOICE OF A MARKETING MIX IN THE SPACE SECTOR.

As described in the previous pages, the number of Marketing Mix instruments seem to have the tendency to rapidly expand, at least in two dimensions

- The number of Ps is expanding
- The subdivision of Ps is equally expanding.

Blois [33] has warned against this over-concentration on 4 (now more) Ps, stressing that different people and different situations require different solutions. Specifically he warns that a rigid concentration on for example the price element in the Social marketing sector may lead to neglecting the prime objective of the organisation. He even argues that a Marketing Mix without the "P" for Price should be possible as well.

As Kotler [9] stated:

> *McCarthy's classification is especially useful from a pedagogical point of view. Nevertheless the feeling remains that some other classification, still to be borne, will develop better conceptual distinctions among the large variety of marketing decision variables.*

Still, he quickly concludes that:

> *The issue is not whether there should be four, six or ten Ps so much as what framework is most helpful in designing marketing strategy. Just as economists use two principal concepts for their framework of analysis, namely demand and supply, the marketer sees the 4Ps as a filing cabinet of tools that could guide their marketing planning.*

Specifically in the nonprofit sector, various authors warn against the excessive use of one of the instruments of the Marketing Mix. Drucker [34] p.53 shows concern about focusing on the financial aspects, expressing this as follows:

> *Napoleon said that there were three things needed to fight a war. The first is money. The second is money. And the third is money. That may be true for war, but it's not true for the nonprofit organization. There you need four things. You need a plan. You need marketing. You need people. And you need money.*

A recent study of Warlop et al. [35] reintroduces the social dimension, which was initially introduced by Rotschild [36] arguing that:

> *It is hard to sell brotherhood like soap*

Looking into the problem of waste management and recycling, it becomes evident that the normal marketing instruments (such as "penalties" for inappropriate sorting and public information campaigns) are not sufficient as motivators. An additional, social awareness dimension will be needed to achieve proper results. Also the authors conclude that there is no single strategy which can be developed and applied to all circumstances [35] p. 22:

> *Our analysis (...) suggests that there are different kinds of brotherhood, each with their analogies in traditional consumer marketing.*

As usual, extremes are seldom optimal. Assael [37] comes to the conclusion that the most important element in the marketing mix for nonprofit organisations is promotion, specifically advertising, personal selling and public relations. However, at the same time he warns, [37] p.732:

> *Nonprofit agencies put too much emphasis on advertising and personal selling. They tend to ignore the other elements of the marketing mix.*

Gwin [38] goes even one step further. He feels that a separate approach is even needed within the same organisation, because, in his opinion, the constituent groups are independent from each other. Phrased in the form of a paradigm he defends his opinion as follows:

> *Clearly, resource generators and service users are the two most important groups in nearly every case. However, many not-for-profit organisations focus so much organisational efforts towards those groups, that they strategically ignore the others. (...) The view proposed by this paradigm is that each constituent group served by the not-for-profit organisation requires a separate and distinct strategic approach, based on the needs of the constituent group targeted.*

Summarising these opinions:

- The 4Ps are undisputedly not a "Cure for all Pain"
- The introduction of the 4Ps in the nonprofit sector needs careful analysis
- Other Ps appearing on the "market" are trying to remedy this
- Different groups require different emphasis on the Ps.

It looks as if the only solution is the introduction of a selective Marketing Mix, adapted to each specific case. Using the aforementioned ICNPO classification, a first attempt is made in table 2.3 to come to such selective approach, whereby:

P1 = Product
P2 = Price
P3 = Physical Distribution
P4 = Promotion
P5 = People
P6 = Physical Evidence
P7 = Process
P8 = Philosophy
0 = Low relevance
X = Relevant
XX = Highly relevant

GROUP	P1	P2	P3	P4	P5	P6	P7	P8
1. Culture and Recreation	XX	XX	XX	XX	0	0	0	XX
2. Education and Research	XX	XX	X	XX	XX	X	0	X
3. Health	XX	X	XX	0	XX	X	X	0
4. Social Services	XX	0	X	0	X	0	0	XX
5. Environment	XX	X	X	XX	0	0	0	XX
6. Development and Housing	X	XX	XX	X	0	X	0	0
7. Law Advocacy and Politics	XX	X	X	0	XX	0	0	X
8. Philanthropic/ Volunteers	XX	0	X	X	XX	X	0	XX
9. International	X	0	X	X	0	0	0	X
10. Religion	XX	0	XX	X	X	0	0	XX
11. Professional Organizations	X	0	X	X	X	0	0	X
12. Not Elsewhere Classified	Not Applicable							

Table 2.3: Selective Marketing Mix for nonprofit Organizations

Some preliminary observations are made as follows:

- The Price element indeed seems not always to be relevant
- Some of the new Ps seem to have a relatively low validity, if any. One could even conclude putting them back inside the more traditional Ps as a subset
- In specific cases People have a high relevance (e.g. the reputation of specialists in health care, reputation of professors in education)
- A generally applicable element seems the Philosophy, even being dominant in certain cases such as Religion and Charity Marketing,

In a conclusion from this evaluation for our specific space environment, and in response to the question "Which Marketing Instruments do we use to reach the general public", a pragmatic approach is followed in the next chapters based upon

- The classical 4Ps will be maintained and worked out consecutively in the next chapters
- As it is felt strongly that space activities cannot be "sold" without adding the philosophical dimension, this element will be added as an extra P.

2.7 CONCLUSION

Marketing is now fully accepted in the nonprofit sector on the whole and is professionally handled by most organisations (a number of them even spending in the order of 10% or more of their budget on Marketing activities). In certain areas, where the market was "protected" by regulations and where traditionally there was no competition, this is still not fully the case. However, in general, commercial interests appear that, in combination with deregulation, lead to such competition. Often it takes a long time before the established (public) entities even become aware of such competition and, due to their traditional structures, are able to counter-react or adapt quickly.

For these reasons, it was felt of importance to describe how nonprofit marketing in general originated and to point out the major differences between profit and nonprofit organisations. These differences are also important in order to understand a number of possible criticisms against the use of marketing in the nonprofit sector and public scrutiny.

However, the transfer of marketing approaches from the profit to the nonprofit sector, has faced some inconsistencies. In order to remedy this, nonprofit authors have designed a number of new Marketing Mix instruments, but a first analysis shows that these instruments also are only partially applicable for different cases; some of them even seem to be a subset of the "old" 4Ps.

Therefore, a general approach for the nonprofit sector looks risky, if not doomed to failure. A selective approach whereby the Marketing Mix is adapted to each specific situation seems more appropriate. Nevertheless, whereas such approaches can be considered to give a better and detailed insight, the simple 4Ps approach should not be fully abandoned, also for its memotechnical merits. For this reason, the following chapters will continue to follow the four traditional marketing tools, (Product, Price, Physical distribution and Promotion) but one extra dimension will be added, namely the Philosophical one, which will be elaborated in a separate, additional, chapter.

Emphasis is put on the fact that this approach is specifically valid in a public (nonprofit) driven environment, but above all, that it is in the interest of all space professionals to "market" space activities better to the general public.

REFERENCES CHAPTER 2

1. SMITH, A., *The Wealth of Nations* (Dent and Sons Ltd., Letchword, 1776).
2. McCARTHY, E.J., *Basic Marketing: A Managerial Approach* (R. Irwin, Illinois, 1960).
3. LEUNIS, J., *Inleiding tot de Marketing* (Acco, Leuven, 1998, 8th ed.), p. 17.
4. KOTLER, Ph. and ARMSTRONG, C., *Marketing. An Introduction.* (Prentice Hall, Englewood Cliffs, 3rd ed., 1993) p. 3.
5. CROSIER, K., What Exactly is marketing? *Quarterly Review of Marketing*, (Winter 1975).
6. KOTLER, Ph., *Kotler on Marketing.* (Free Press, 1999) p.121.
7. LEVITT, T., Marketing Myopia. *Harvard Business Review*, Vol. 38 (July-August, 1960), pp. 45-56.
8. KOTLER, Ph. and FOX, K., *Strategic Marketing for Educational Institutions.* (Prentice Hall, Englewood Cliffs, 1985), p. 5.
9. KOTLER, Ph. and ANDREASEN, R., *Strategic Marketing for Nonprofit Organizations* (Prentice Hall, New Jersey, 5th ed., 1996)
10. ZEITHAML, V. and BITNER, M.J., *Services Marketing.* (McGraw-Hill, Singapore, 1996)
11. LEVITT, T., Marketing intangible products and product intangibles. *Harvard Business Review*, Vol. 59 (March-June 1981) pp. 94-102.
12. KOTLER, Ph. and LEVY S., Broadening the Concept of Marketing. *Journal of Marketing*, Vol. 33 (January 1969), pp. 10-15.
13. SHAPIRO, B., Marketing for nonprofit organizations. *Harvard Business Review*, Vol.51 (September-October 1973), pp. 123-132.
14. HUNT, S., The Three Dichotomies Model of marketing: An Elaboration of Issues. In *Macro-Marketing*, SLATER, C. (University of Colorado, Boulder, 1977) pp. 52-56.
15. KOTLER, Ph., Strategies for Introducing Marketing into Nonprofit Organizations. *Journal of Marketing*, Vol. 43 (January 1979), pp. 37-44.
16. SARGEANT, A., *Marketing Management for Nonprofit Organizations.* (Oxford University Press, Oxford, 1999), p. 17.
17. HANSMANN, H., The Role of Nonprofit Enterprises. *The Yale Law Journal*, (April 1980), pp. 835-901.
18. LOVELOCK, C. and WEINBERG, C., *Marketing for Public and Nonprofit Managers*, (J. Wiley & Sons, New York, 1984) p.31.
19. SALAMON, L. and ANHEIER, H., *The Emerging Nonprofit Sector.* (Manchester University Press, Manchester, 1996).
20. ANTHONY, R.N., and HERZLINGER, R.E., *Management Control in Nonprofit Organizations.* (Irwin, Homewood Illinois, 1980).
21. WEBER, M., *The Theory of Social and Economic Organizations*, originally published in German (1922) (English translation: Free Press, New York, 1957).
22. DRUCKER, P., *Discontinuity*, (Harper&Row, New York, 1969) p.229.
23. LOVELOCK, C. and WEINBERG C., Public & Nonprofit Marketing, (Scientific Press, San Francisco, 2nd ed., 1990) pp. 4-7.
24. ANDREASEN, A., *Marketing Social Change.* (Jossey-Bass Publ., Washington, 1995).
25. DRUCKER, P., Managing the Public Service Institution. *The Public Interest*, Vol. 33 (1973), pp. 43-60.
26. BORDEN, N., The Concept of the Marketing Mix. *Journal of Advertising Research*, Vol. 4 (June 1964), pp.2-7.
27. LIPSON, H. and DARLING, J., *Introduction to Marketing: An Administrative Approach*, (J. Wiley, New York, 1971) pp. 585-621.

28. DOYLE, P., Managing the Marketing Mix. In M.J. Baker (ed.), *The Marketing Book*, (Heinemann, London, 2nd ed. 1991)
29. BLOOM, P, and NOVELLI, W., Problems Applying Conventional Wisdom to Social Marketing Programs. In M.P. MOKWA and S.E. PERMUT, *Government Marketing. Theory and Practice.* (Praeger, New York, 1981).
30. BOOMS, B. and BITNER, M., Marketing Strategies and Organisation Structure for Service Firms. In DONNELY, J. *Marketing of Services.* (AMA, Chicago, 1981).
31. BRUCE, I., *Successful Charity Marketing*, (Prentice Hall, Hertfordshire, 2nd ed. 1998)
32. EMBLEY, L., *Doing Well While Doing Good.* (Prentice Hall, Englewood Cliffs NY, 1993)
33. BLOIS, K., Marketing for Non-profit Organisations. In M.J. BAKER, *The Marketing Book,* (Heinemann, London, 1987)
34. DRUCKER, P., *Managing the Non-profit Organization. Principles and Practices.*(Harper, New York, 1992)
35. WARLOP, L., SMEESTERS, D., VANDEN ABEELE, P., *Selling Brotherhood like Soap: Influencing Everyday Disposal Decisions.* (KUL Research Report 9952, Leuven, 1999)
36. ROTSCHILD, M., Marketing communications in nonbusiness situations or why it's so hard to sell brotherhood like soap. *Journal of Marketing.* Vol.43 (Spring 1979) pp. 11-20.
37. ASSAEL, H., *Marketing Management. Strategy and Action.* (Kent Publishing, Boston, 1985)
38. GWIN J., Constituent Analysis: A Paradigm for Marketing Effectiveness in the Not-for-profit Organisation. *European Journal of Marketing*, Vol.24/7 (1990) pp. 43-48.

Chapter 3

THE SPACE PRODUCT

3.1   INTRODUCTION

In this chapter emphasis will be placed deliberately on the marketing related product aspects and not on the space product design itself. There are two rationales which have led to this choice:

1. There are a number of excellent books available, which provide full course texts on various space products. Many of these books are based upon years of lecturing by specialists and it would make no sense to try to copy extracts of such books here. It is our conviction that it is preferable to make direct reference to the following books:

    - "Space Mission Analysis and Design" [1], which gives a generic overview of all aspects related to space products in general,
    - "Human Spaceflight" [2], providing a similar approach, but with emphasis on the problems associated with manned spaceflight.
    - "Space Stations" [3], where the emphasis is put on space stations in general and the associated elements such as logistics and utilisation, whilst at the same time giving an overview of ISS in particular.

2. Such detailed descriptions of the product would physically bring the Product, as one of the tools from the Marketing Mix, completely out of proportion. It is the objective of this work to show that a balanced set of tools are needed to reach a global marketing objective and we will concentrate only on the role of the product as a marketing tool in space activities.

In a first instance, we will try to analyse why products in the nonprofit sector in general and space products in particular have certain specific differences compared to other industrial and commercial products. However, in analogy with products in any other sector, this means that we will investigate the choice of space products/projects and how such choice can contribute to the efficient use of the resources allocated to the organisation.

The "Good news is no news" syndrome is very strong in a sector which has a considerable number of critics. Therefore failures in space projects lead to very strong critical reactions and cannot be ignored in this marketing concept. Reliability will be considered from this perspective. However, probably more important than the direct product is the spin-off product, as an argument in favour of space activities. This is not due only to the objective values, where one can prove that space activities have a very positive multiplier effect, but also due to the tangible effect of every-day products whereby the general public comes in touch physically with the spin-off of space activities.

3.2 SPECIFICS OF SPACE PRODUCTS

PRODUCT IN NONPROFIT MARKETING

Many authors warn about the "unfiltered" transfer of management techniques from the private to public sector, specifically in the field of product marketing tools. Micklethwait and Woolridge [4] warn against the introduction of management guru's (which they attribute the name "Witch Doctors") in government practices, as was considered innovative in the 1980's and early 1990's (especially under the Clinton/Gore administration). The product provided by the nonprofit organisation has an economic aspect, to which market economy and traditional management techniques are applicable (e.g. the price of health care or the price of public transport). However, there is also a service dimension, whereby governments undertake to levy the services, not in accordance with the income (taxes paid), but in accordance with the need (e.g. free medical care for certain social categories, reduced or free public transport for pensioners etc.). Certainly a service needs to be efficient, but it remains above all a service.

One example quoted in [4] p.320 are the so-called Californian "contract cities", where most of the public sector services were bought in. Lakewood, a city of 70,000 inhabitants, started off with only 3 employees to monitor the services, which were contracted out (including hospitals, fire brigade and even the police). However, in 1995 the number of civil servants was back up to 160 and the Council was meeting every two weeks, rather than, as originally planned, once a year. Even when the per capita spending on public services proved indeed to be only half, the population strongly preferred to pay more for ... service. The authors quote H. Minzberg [4] p.327 to even have stated at a conference on the influence of management theory on the public sector:

> ***The Harvard Business Review should have a skull and crossbones stamped at the cover with the warning "not to be taken by the public sector"***

Two aspects have a generic interest in literature in nonprofit marketing, with respect to the product:

1. Also nonprofit organisations need to have contact with the customer.

Andreasen [5] summarises a number of potential problems, which occur in nonprofit organisations in this respect; namely:
- The offering is seen as desirable anyway, without the need for justification
- The consumer is considered to be ignorant
- Overemphasis is placed on promotion and PR
- Consumer research is considered of secondary importance
- Marketing staff is selected on the basis of product knowledge.

2. Performance and efficiency are important aspects in the nonprofit sector

Management experts such as Peter Drucker constantly remind us that the prime customer of public services is the public itself. The point, according to Drucker, is not how to shrink or save but how to be more effective. He expresses this [6] p. 107 as:

> ***Non-profit institutions tend not to give priority to performance and results. Yet performance and results are far more important – and far more difficult to measure and control – in the non-profit institution than in business.***

The aspects of performance and performance measurement will be put in the context of space activities, later in the text. But first, we have to make a distinction between a number of factors in this chapter. We could tentatively group them into:

- **The direct product**: i.e. the satellite, launcher, etc. respecting a number of requirements and fulfilling the primary role it has been designed for.
- **The spin-off**: Due to the unique environmental conditions and the limitations in weight (upload), a large range of more or less known derivatives have originated from space activities. Moreover, specific techniques and systems have been widely used in other sectors.
- **Non-tangible product**: although this always existed, there was a certain reluctance to defend space activities with non-technical arguments. More recently, fortunately, this attitude has changed and one discusses more openly the political and philosophical aspects. This aspect, from a marketing point of view, is considered so important that it will be treated in a separate chapter.

THE DIRECT SPACE PRODUCT

Any object designed and developed to perform a function in space has to withstand a very hostile environment during its designed lifetime. On top of this, conditions during launch give another set of acoustic, vibration and acceleration constraints, which can prematurely prevent proper functioning. Of course this is a highly visible factor, and therefore extreme care is taken in testing the models in advance and building in redundancy. This is probably also the reason why in an earlier study Bloomquist [7], after having analysed 374 spacecraft, found out that on the average not less than 2.8 times the design life was obtained. Furthermore, research performed on 2500 spacecraft failures in a similar period (1962-1988), shows us that still a considerable portion of such failures can be assigned to design and operational causes, as per table 3.1, deduced from [1].

CAUSE OF FAILURES	OCCURRENCE (IN %)
Design	24.8%
Environment	21.4%
Operations	4.7%
Parts	16.3%
Quality	7.7%
Other (known)	6.3%
Unknown	18.9%

Table 3.1: Cause of spacecraft failures

Product Assurance methods have a long-standing "tradition" in space activities. In Europe, ESA has taken the initiative of the ECSS (European Cooperation on Space Standardisation) and produced standards, which are now generally applied in European space industry. Information and lists of these standards can be found under

http://industry.esa.int/

Some of these standards have been transferred to other industrial sectors and are extensively applied there (one of the examples of a non-tangible space spin-off). An excellent example are the "Software Engineering Standards", which are presently commercially published and have sold over more than 8000 copies. These originated as ESA software engineering standards under the ECSS initiative.

The fact that a non-negligible number of system failures were recorded has led to the introduction of Total Quality Methods (TQM). W. E. Deming introduced the basis for this approach, after extensively studying Japanese Quality Circles. The resulting work "Out of the Crisis" [8] was at its 22nd edition in 1994, which certainly illustrates the success of the book. The 14 initial points are summarised in [9] and presented in table 3.2

DEMING'S 14 POINTS FOR TQM

1. Create constancy of purpose towards improvement of product and service.
2. Adopt the new philosophy. Do not longer live with commonly accepted levels of delays and mistakes.
3. Cease dependence on inspection. Require, instead, statistical evidence that quality is built in
4. End the practice of awarding business only on the basis of the price tag.
5. Find problems. It's the management's job to work continually on the system.
6. Institute modern methods of training on the job.
7. Institute modern methods of supervision.
8. Drive out fear, so that everyone may work effectively for the company
9. Break down barriers between departments.
10. Eliminate numerical goals, posters and slogans
11. Eliminate work standards that prescribe numerical quotas
12. Remove barriers between the hourly worker and his pride for workmanship
13. Institute a vigorous program of education and training
14. Create a structure in management that will push every day on the above thirteen points.

Table 3.2: Basic TQM rules according to Deming

A number of case studies on how TQM can be used in the nonprofit sector are described in [10], whereby a more practical approach is given as per following steps:

1. State the project objectives
2. Define the current stage
3. Define the desired stage
4. Identify the gaps between the current and the desired stage
5. Develop a competitive analysis process
6. Implement, perform training, and evaluate the competitive process.

We can illustrate this process by referring to a European study on Space Exploration and Utilisation, summarised in [11]:

- Step1: Taking into account the political, financial and industrial parameters, a shortlist of future European Space scenarios has been made. From this analysis, Mars exploration, Moon exploration, Space Solar Power and Space Tourism were selected as possible candidates
- Step2: For each of these cases, an "inventory" was made of the present European capacities to reach these objectives, specifically in terms of technological experience
- Step 3: Looking backward, an analysis was made on which technologies would be needed to reach the objectives, again putting the emphasis on technology
- Step 4: Comparison of both previous steps leads to a number of "Technology Roadmaps" for each of the four selected scenarios which describe when the lacking technological developments are needed to be timely available
- Step 5: For each of the cases, sharing of interest and partnerships are suggested for the development of these technologies.

This brings us to the most critical product factor at present, the launcher. In the late eighties, launch insurance premiums reached levels, which were virtually not affordable anymore, at a certain point being in the range of 20-25%. It is evident that under these circumstances satellite operators looked primarily at the reliability rates of the commercial launchers to make a decision in this area. Ariane could then show, in the early 90's, a success rate of 93%, versus e.g. General Dynamics 80% and Atlas only some 73%. (In all fairness one must admit that the Delta rocket then could demonstrate a rate of not less than 98.6 % over 14 years). Coupled with interesting commercial conditions, this led to a considerable influx of new orders for Ariane, obtaining some 50% of the commercial launch market.

THE SPACE PRODUCT MARKET

Typical for the space market is that one and the same product (e.g. a transponder) can be marketed in a number of markets with very different boundary conditions. A first important distinction is the budget-driven markets versus the industrial markets, as shown in figure 3.1 (the industrial markets are marked in the figure in darker grey, the budget-driven ones in lighter grey)

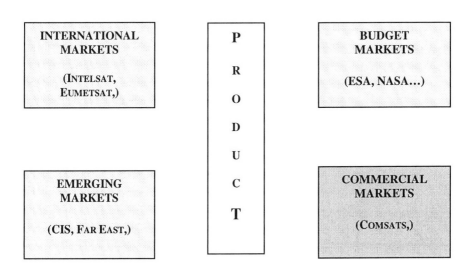

Fig. 3.1 Overview of the space markets

There are major differences between the budget-driven markets and the International markets; the latter ones having income which (partially) covers the expenses.

The International markets have clear defined objectives and clearly defined Life Cycles in their products; most of the time the next generations are known well in advance. (Example: Meteosat Second Generation, which has to be ready at a certain date as the first generation reaches its end of designed lifetime). The next generations may be postponed in time, due to problems with the financial and political approval cycles, but the final product will not. Globally speaking, there is certain stability in the International markets due to this continuity and a fairly predictable turnover for industry.

The pure Budget sharing markets are more susceptible to changes in the environmental conditions. Changes in the macroeconomic environment or political environment, and even changes in Government may give raise to a sudden switch in priorities. One of the problems associated with this are the long term utilisation scenarios (e.g. Eureca, Spacelab) when the policy makers want new, visible products rather then supporting the utilisation of products designed by previous policy makers. Typical examples in the European space sector are the cancellation of the Hermes project (changes in French policy) and the cancellation of joint military projects in the Earth Observation era.

Therefore, the predictability of these markets is more risky for industry. From a business policy point of view industry should make sure that only a certain percentage of the turnover, for these reasons, is put on the Budget sharing markets.

A considerable difference for the producer is also the freedom of design as follows:

- In the open market he can make is own design and adapt it if needed
- In governmental markets, specifications are very accurate (e.g. quality standards)
- In a number of markets, there are additional restrictions (e.g. geographical return).

3.3   THE PRODUCT CHOICE

In any system, a limited number of new products can be regularly brought on the market. This restriction is driven by limitations in resources and the associated costs required in order to introduce such new products. A similar effect can be seen in Agency projects. Budget restrictions lead to the effect that there are always more ideas and projects than can be executed within the budgetary resources. Therefore in both cases it is important that the right choice is made on the product or project that will be initiated. Whereas for a product this success can be measured relatively quickly in terms of sales, "success" is more difficult to measure in the nonprofit sector. Research done in the field of NPD (New Product Development, under which term this area is known in Marketing) has led to a number of conclusions which are useful for space activities as well, and therefore will be briefly described hereafter.

The first extensive study in this field was the "SAPPHO" (Scientific Activity Predictor from Patterns with Heuristic Origin) project, led by the noted economist Christopher Freeman. The project was undertaken in the 1970s at the Science Policy Research Unit (SPRU) and concentrated on innovations in the chemical and instrumentation industry (it is interested to note that other authors in that period also considered the chemical industry as one of the most progressive business innovators).

Examining such innovative products, more than 200 measures of innovative aspects were used, from which, after analysis, only 15 were retained as statistically relevant. By far the first determining factor was "Successful firms understand user needs better", which led to the main conclusion [12] p.197:

> ***Successful innovators innovate in response to market needs, involve potential users in the development of the innovation, and understand the user needs better.***

Cooper [13] examined a random sample of 100 successful and 100 failed products more from a marketing perspective, and correlated them to a number of variables. His conclusions include (the factors are given in order of importance):

Three keys to success:
- Product Uniqueness and Superiority
- Market Knowledge and Marketing Proficiency
- Technical and Production Synergy.

Three facilitators for success:
- Marketing and management strategy
- Strength of marketing communications
- Market need, growth and size.

Three barriers for success:
- Product too high-priced, compared to competitors
- Market very dynamic
- Competitive market, with "satisfied" customers.

The results of these and similar investigations became better known when the book "In Search of Excellence" [12] was published (from which more than 2.5 million copies were sold in the early 1980's, putting it high on the Best-seller list). The authors group the different findings as eight basic principles:

1. Create a bias for action (similar to the "don't think, do" statement)
2. Stay close to the customer, learning his preferences
3. Encourage internal autonomy and entrepreneurship, even up to the point of breaking the corporation in smaller companies
4. Productivity comes through people, create awareness thereof by all employees
5. Propagate "hands-on, value driven", whereby executives should stay in touch with the firm's essential business
6. Stick to the knitting; remain in the field you know best
7. Create simple forms and lean management structures
8. Maintain simultaneous loose-tight properties, combining dedication and discipline with tolerance.

The authors put a lot of emphasis on the need to build up a close link with the users, also because many successful products and ideas come from those users. An extensive study on scientific instruments revealed that [12] p.194:

- Of the 11 "first of type" instruments, all came from users (including the gas chromatograph, nuclear resonance spectrometer and transmission electron microscope)
- Of the 66 "major improvements", 85% came from users
- Of the 83 "minor improvements", two-thirds came from users.

In order to verify the validity of the findings in other cultural environment, similar questions and/or the same questionnaires were applied in other countries as well. In general, the findings were found to be globally valid, be it with minor differences. In Belgium, for example, it was found that smaller firms were rather reacting to a "Market pull", whereas larger firms were more oriented towards a "Technology push", whereby the aggressiveness of its marketing tools determine the extent of success in this latter case [14].

A number of studies and research results in this NPD field are published in the "Journal of Product Innovation Management". Here we also find a number of quantitative tools to measure R&D effectiveness such as the R&D EI (Effectiveness Index) defined as [15]:

$$EI = \frac{\% \text{ New Product Review} \times (\text{Net Profit \%} + \text{R\&D \%})}{\text{R\&D \%}}$$

Therefore, if a firm has 40% of its revenue from new products, a net profit of 9% and a research investment equal to 6% of its revenue, the index would be 1 (Computation: 40% x (9% + 6 %) / 6%). When the index is above 1, the return from new products is running at a greater rate than the investments.

Here again, we encounter the problem in the nonprofit area that there is no quantitative measurement in view of the lack of profit measurement (although this can be overcome by other techniques, as will be discussed later). One space related case reported concerns SPACEHAB [16], which according to the authors scores high w.r.t. its closeness to the users and understanding of the market. Analysis provides following characteristics, as summarised in table 3.3:

PROTOCOL DIMENSION	SPACEHAB CHARACTERISTICS
Performance advances	Provides additional critical crew space, a major increase in Orbiter capabilities and payload integration
Customer interfaces	Builds on NASA/customer relationship to serve their needs better
Need fulfilment	Responds to user needs for payload requirements, manifest, and flight support
Benefits	Provides faster experiment integration cycles, lower cost per experiment, and augments existing initiatives

Table 3.3: User Relations SPACEHAB

For the nonprofit type of activities, such analysis is more difficult because the direct effect is not easily measurable and often, even the final customer is not clearly defined.

We can illustrate this latter aspect with a survey made in the Dutch Civil Services [18]. Overall, 40 % of higher civil servants considered their minister as the customer (in the case of the Ministry of Interior, this was even 60 %). In the same review, 27 % of the respondents agreed that they were performing services, which were not really useful for the general public.

In the case of NASA, the so-called Augustine Committee concluded in December 1990 that "the goals put by NASA were not reflecting the expectancies of the general public". In August 1991, a U.S. House Subcommittee opened an investigation into NASA's management practices "to learn why the Agency that for so long epitomised excellence has lost the claim to that honour". The report was called "NASA's Midlife Crisis" and accelerated important changes. The newly appointed NASA administrator, Dan Goldin, consequently stated that "no national goals will be supported over the long term unless the public is motivated."

One of the successful achievements in this respect is ESA's scientific programme. Here a good interaction is maintained with the scientific user community and a number of projects are regularly proposed within a certain framework (e.g. Horizon 2000). The scientific objectives are set in advance and regularly a project is added which satisfies one of the objectives, gradually changing from one topic to another over time. This allows the scientists to plan the development of their payloads well in advance and guarantees a good relationship between the provider (ESA) and the customer (the European space science community), who feel closely involved in the decision process [17].

Recent studies also emphasise the role of the product itself in the communication process with the customer. This is based upon the observation that [18]:

- Consumers retain only 20% of what they see and 30% of what they hear
- Of the 925 commercial broadcasts shown on Dutch TV in 1996, only 20 are actually retained
- In 1989, these figures were respectively 100 and 7, so the percentage decreases with increasing information.
- Consumers, however, retain up to 80% of what they saw, heard and physically got in contact with.

Therefore, it is important to bring the potential "customer" in contact with the product as early as possible, even if this is only in the design or prototype stage. (Architects know this aspect very well and develop scale models of projects very early as promotion tools). The situation has improved with modern techniques such as virtual reality, and modern research strongly suggests ensuring an early physical confrontation with the product by the customer (such as prototype models after phase A studies or access to virtual reality demonstrations, eventually distributed via Internet).

3.4 EFFICIENCY IN SPACE PRODUCTS

One cannot stress sufficiently the difference between a budget-driven organisation and a profit-making, cost driven organisation. If we express this in terms of objective functions, a profit-making organisation tries to maximise the difference between income and expenses (hence the profit), whereas a nonprofit organisation tries to minimise the difference between expenses and budget, by staying within the budget.

A nonprofit organisation has a given budget and its aim is to give a maximum return, to a value requested by his customers and within that budget. Therefore a nonprofit organisation does not calculate in terms of Return On Investment (ROI) but in terms of budget compliance. Hence, any marketing concept, which does not comply with this budget structure, is doomed to fail. In the case of space activities, one should not hope also for "deus ex machina" increases in the budget. Major space participating countries work with Mid-term revolving budgets (typically 5 years) and therefore it is possible to predict the budgets of the upcoming years rather accurately. By no means should the lack of profit result in a lack of control over expenditure; the only big difference is the problem of how to measure the efficiency.

EFFICIENCY MEASUREMENT IN THE NONPROFIT SECTOR

In the traditional organisations profit centres are created. It can be relatively objective to determine profit per area/department/division etc. at the end of the accounting period, depending on how analytical bookkeeping is carried out. This is not the case for the nonprofit sector. Did the responsible person from a certain area who stayed within the budget, perform an efficient job or did he simply spend the money? The key to this question, in general terms, is to find an objective parameter that can be measured. It is typical to note in this context that one of the first statements of Dan Goldin, the NASA administrator, was:

If you can't measure it, you can't manage it

This is a typical example of the introduction of "traditional" management techniques in the nonprofit sector (Goldin was a TRW-director before being appointed at NASA in 1992).

If we go back to the real definitions, we see that it is not so difficult to find tangible parameters also in the nonprofit sector. Based upon systems approaches, In 't Veld [19] defines:

- Effectivity = the ratio between real output and expected output
- Efficiency = the ratio between real inputs and expected inputs
- Productivity = the ratio between real output and real input.

Or, expressed in formulae with:

O_R = Real Output (i.e. the measured one)
O_E = Expected Output (i.e. the one assumed)
I_R = Real Input (i.e. the measured one)
I_E = Expected Input (i.e. the one assumed)

- **Effectivity = O_R / O_E**
- **Efficiency = I_R / I_E**
- **Productivity = O_R / I_R.**

Under these definitions it is possible to find key indicators in any organisation, both on the input (e.g. resources) as on the output side and to measure not only efficiency, but also even effectivity and productivity.

The use of such quantified measures is known as the "Management by Objectives" method (MBO). This method was initiated in the early 70's for U.S. federal programmes [20] and was gradually implemented in other nonprofit sectors as well.

MBO is a formal procedure that stresses goals and outputs and depends for its success on feedback and performance reports. The basis is to express objectives in quantified terms. Examples are:
- Number of files processed
- Number of drawings produced
- Attendance figures

CHAPTER 3 : THE SPACE PRODUCT

In general terms, any MBO type of process follows four steps:

- Step 1: Goals Setting (What are the targets?)
- Step 2: Performance Measurement (What are the results?)
- Step 3: Performance Analysis (What are the differences?)
- Step 4: Corrective Action (What shall we do about it?).

In order to bring the changed philosophy to general attention, Goldin introduced in NASA a new concept in 1992, which became widely known as the "Faster, Cheaper, Better" approach. On the restructuring side, Goldin presented results in 1998 [21], which indeed look impressive, and are summarised in table 3.4

CATEGORY	FY 1993	DECEMBER 1997	FUTURE (2000+)
Civil servants	24,900	19,187	17,818
Supervisor Ratio	5.4:1	9.6:1	11:1
HQ Staff	1,344	1,022	954

Table 3.4: NASA restructuring results

More important were the results directly related to the concept presented at the same occasion and summarised in table 3.5.

CATEGORY	FY 90-94	FY 95-99	FY 00-04
Average Development (yrs)	8.3	4.4	3.5
Annual flight rate	2	9	13
Average Cost (1995 million $)	590	175	85

Table 3.5 Faster, Better, Cheaper evaluation

A study, made by Aerospace Corp. [22] partially confirms this tendency (the study was based upon 28 completed NASA science missions). Indeed, the resulting figures from the study were:

- Reduction in time from 6.6 to 3.25 years
- Average flight rate increased from 1.25 to 2.75 flights per year
- Average cost reduced from 654 million $ to 96 million $.

However, the study also revealed (Goldin did not present comparable figures) that:

- Catastrophic failure rate went up from 10 to 28%
- Partial failure rate went up from 30 to 44%
- Scientific use reduced from 305 "instrument months" per mission to 79.

Although in pure economical terms this means that the new approach came to a cost effectiveness factor of 0.82 (instrument months divided by the cost), compared to previously 0.52, the authors put question marks on whether this is publicly a valid argument. A failure like the 1 billion $ Mars Observer one has caused a considerable loss of reputation for NASA which cannot be compensated by the efficiency gain.

Also in ESA, the trend to introduce a number of Performance Indicators can be noted. This has to be seen in the context of growing criticism on efficiency, which ESA tries to counter, by a higher level of transparency. Based upon the previously mentioned emphasis on time and cost reduction, the choice of indicators reflects an emphasis on quality of the product. Indicators, as agreed upon in 1998 cover:

A set of pure performance indicators, e.g.

- Number of technology achievements
- Number of new proposals
- Development time
- Amount of data recovered per spacecraft
- Percentage of data utilised.

A set of cost efficiency indicators

- Rate of ESA non-programme costs
- Rate of ESA staff costs
- Staff cost efficiency.

As well as a set of pure financial indicators

3.5 THE SPACE PRODUCT DEVELOPMENT CYCLE

RELIABILITY

If we talk about reliability we refer mainly to mission reliability, which stands for the probability that the essential mission elements, such as those related to safety, will survive (as opposed to basic reliability, where all elements have to survive).

One of the prime methods to provide such reliability is the use of redundancy, whereby a second (or a third) system can immediately take over if one fails (e.g. different computers on board of the Shuttle with different programming to avoid bugs).

The most visible failures for the general public are evidently launch failures. The situation has largely improved over time, progressing from a reliability of some 80% to reliability, at present of 96% (Ariane 4, based upon 79 launches). An overview of launch failures, as published in [22] is provided in table 3.6

LAUNCHERS	1968/1977	1978/1987	1988/1997
U.S.	9.1 %	6.1 %	5.4 %
CIS[1]	6.9 %	3.6 %	4.7 %
Europe[2]	100 %	20 %	4.8 %
China	15.4 %	15.2 %	17.9 %
Japan	0 %	5.6 %	12.1 %

Table 3.6: Evolution of launch failures.

Notes:

[1] Launches serving manned programmes have considerably higher reliabilities. The Russian Progress rocket, as an example, has a launch record of 84 consecutive successful launches (status 9/2000, source : ZPK).

[2] Refers to European cooperative programmes.

92 SPACE MARKETING

(Photo: ESA-CNES-Arianespace /Service Optique CSG.)

Fig. 3.2: Launch of Ariane 4

As described in another chapter, insurers count on a failure rate of less than 4% for Ariane5 and are willing to insure the associated risks. Such failure rates may be acceptable for unmanned launch systems; but to the public they are not acceptable for manned ones.

One cannot avoid touching the Challenger accident in the context of reliability and safety issues, when covering manned space activities. A very important aspect was found during the investigation of this tragic accident by the so-called Rogers Commission and is reported in [24] p.209, as a case history. Bearing in mind that the accident occurred on 28 January 1986, it is surprising to note the following chronological events:

- January 1979: The Chief of the Solid Motor Branch informed management on the malfunctioning of the O-ring seal

- May 1980: A NASA engineering panel noted that the O-ring seals failed during a ground test

- December 1982: The Engineering Panel added the O-ring to the criticality list with the remark that "it lacked a reliable backup and, if the joint failed, it would lead to a loss of mission and crew"

- January 1985: Thiokol engineers reported extensive damage on the O-rings after a launch with low overnight temperatures

- 27 January 1986: In a teleconference top management voted against delaying the launch in view of the low temperatures (overruling the advice of engineers).

This unfortunate case study taught us a lot on "risk communication". It is not sufficient to make risk assessments and to install a proper risk management, it is equally important that the messages are not filtered and overruled by promotional or political considerations (there was in this particular case a lot of interest in the first school lesson from space, which had attracted considerable media attention). As a consequence of this, the Rogers Commission recommended "that anyone in the NASA system holding a strong view regarding a safety issue must be permitted to express their opinion at any level".

In the framework of a marketing mix this example again emphasises a basic principle: even if there is a paramount interest in one of the tools, there is no reason to ignore the other components.

Increasing reliability can be obtained by adding redundant systems, often resulting from a different design or approach. The cost/reliability function is an exponential one, due to the inclusion of such redundancies and with higher standards applied built-in to reach higher reliability levels. This effect can be noted at all levels, as can be seen in figure 3.3, where the subsystem development costs for similar systems are compared, respectively for manned spaceflight, unmanned spaceflight and commercial aircraft. [2] p.591.

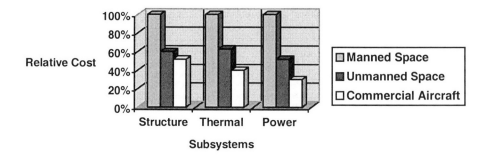

Fig. 3.3 Relative Cost in function of design culture

In the context of reliability, a popular approach in Services Marketing may be applicable here, namely

Doing the Service Right the first time, but Doing the Service Very Right the second time...

Research shows us that in general a first failure is accepted, but that the "zone of tolerance" reduces considerably for the second attempt. A number of suggestions are provided for such "Recovery effort", [25] p.58

- Give an explanation on how the problem happened
- Give immediate feedback when the problem is resolved
- Tell how long it will take to solve the problem
- Give immediately alternatives if the problem cannot be solved
- Explain how the problem will be avoided in the future
- Give progress reports if the problem cannot be solved immediately.

Further research even shows that this way of professional "Recovery" can even increase confidence on the long run. The way the first Ariane5 (A501) failure was handled, in line with the previous guidelines, confirms the validity of this concept.

MEASURING QUALITY

There are, however, more direct techniques, which are used in service marketing to measure the results and the opinion of the public. In the service sector, quality is often measured using the now generally accepted SERVQUAL approach. This approach measures the intangible elements of quality in service and assigns indicators to a set of scales, reflecting the service quality as perceived by the consumers of the service.

The most widely-spread version of SERVQUAL uses a 22-question scale [26]. The questions, as represented in table 3.7, are answered on a seven-point scale, ranging from "Strongly Agree" (rating 7) to "Strongly Disagree" (rating 1) with no verbal labels for the intermediate scale points. This standardisation is important in view of benchmarking, when the performance of different organisations is compared with each other.

Applying this questionnaire to the nonprofit sector, analysis revealed [27] that there was a very high level of convergence in the expectancies in the nonprofit sector, compared with results from the profit-making service sector.

SERVQUAL was applied in a sample of 218 major donors of nonprofit organisations, in order to evaluate the relationship between satisfaction of the donors and their gifts. Again, this not only proved that techniques from the "classical" marketing areas can be transferred from the profit to the nonprofit sector, but also that customers expect the nonprofit providers to meet the same standards and quality as commercial service providers.

A word of caution has to be added to this. Applying the SERVQUAL technique to an hospital environment gave less satisfactory results than expected [28]. Using the standard questionnaire for a group of patients, the many questions led to a degree of confusion and many inconsistencies. The uniqueness of certain services may lead to a need for tailored questionnaires, even to the detriment of losing benchmarking possibilities.

SERVQUAL QUESTIONS

1. They should have up-to-date equipment.
2. Their physical facilities should be visually appealing.
3. Their employees should be well dressed and neat.
4. The appearance of the physical facilities should be in keeping with the type of services provided.
5. When they promise to do something by a certain time. They should do so.
6. When customers have problems, they should be sympathetic and reassuring.
7. They should be dependable.
8. They should provide their services at a time they promise to do so.
9. They should keep their records accurately.
10. They shouldn't be expected to tell customers exactly when services will be performed (*).
11. It is not realistic for customers to expect prompt services from their employees (*).
12. Their employees don't always have to be willing to help customers (*).
13. It is okay if they are too busy to respond to customer requests promptly (*).
14. Customers should be able to trust their employees.
15. Customers should be able to feel safe in their transactions with their employees.
16. Their employees should be polite.
17. Their employees should get adequate support to do their jobs well.
18. They should not be expected to give customers individual attention (*).
19. It is unrealistic to expect employees to give customers personal attention (*).
20. It is unrealistic to expect employees to know what the needs of the customers are (*).
21. It is unrealistic to expect from them to have their customer's best interests at heart (*).
22. They shouldn't be expected to have opening hours convenient to their customers (*).

(*) : reverse-scored ratings

Table 3.7: List of SERVQUAL questions

PUBLIC RESPONSE TO QUALITY

In order to understand how the public perceives the quality of a non-tangible product such as space activities (as they are seldom confronted with the physical evidence of the results), models developed in the area of environmental psychology can be used. One of the frequently used approaches for such cases is the SOR-model (Stimulus-organism-response) [29]. As per figure 3.4, we can distinguish 3 aspects:

- A set of stimuli
- An organism component
- A set of responses

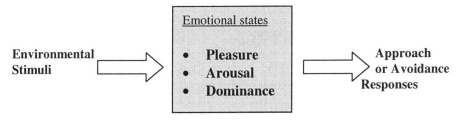

Fig. 3.4: SOR – model

Every service organisation sends out a number of "stimuli" to the environment, such as physical evidence, presentations, leaflets, and press coverage. The target audience can react to this in three different ways:

- **Pleasure/displeasure**: as a typical example we can refer to a launch event, where the viewer will be pleased to witness a lift-off, and displeased in case of a launch delay
- **Arousal/nonarousal**: this refers to a higher state of excitement. Clearly the first pictures in 1969 of the first man on the moon led to such a reaction, whereas the pictures of the Challenger accident worked in the opposite direction
- **Dominance/submissiveness**: reflects the feeling to influence control over the organisation or the feeling that the organisation is well managed. An example could be "letters to the editor" or reactions via politicians.

Based upon these feelings, the target audience will react with a certain "response", such as, in the case of space activities:

- A desire to support (approach) or refusal to support (avoid) space activities
- A desire to get more information (approach) or to ignore (avoid) it
- A desire to communicate (approach) or to resist all attempts of communication (avoid) with the organisation
- Feelings of satisfaction (approach) or disappointment (avoidance) with the results.

This process will take some time and therefore, contrary to traditional marketing, the results of stimuli cannot be measured quickly. However, an understanding of the process is important because, in general, tendencies cannot be easily influenced in the short-term and can accumulate towards for example election periods, where both (or more) parties have to take an a priori position towards space programmes.

Frequently, one of the methods used is a poll, whereby specialised organisations ask regularly or irregularly voters about their opinions. One of the most commonly known polls is the "Gallup polls", that also emphasise demographic differences, which are important input for marketing strategies [30]. In order to illustrate the differences in terms of socio-demographic characteristics, some results of the 1989 Gallup survey are presented in figure 3.5.

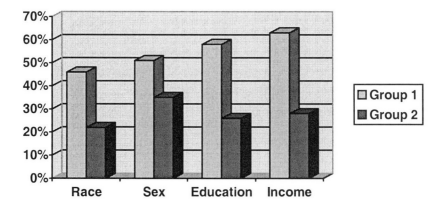

Fig 3.5: Difference in space support (Gallup poll 1989)

In figure 3.5 the following data are graphically represented:

- 46% of white respondents (group 1) were in favour of space activities spending, compared to 22% respondents of other ethnic origin
- 51% of men (group 1) were supportive, versus 35% of women
- 58% of college graduates (group 1) support space programmes, compared to only 26% of those without high school diploma
- 63% of the higher income group (then >50,000$) supported space, versus 28% of the lower income group (then <20,000$).

Although the absolute figures change over time and differ per country, this gives a good indication where efforts should be placed on external stimuli in a cost-effective way. Indeed, to increase for example the number of supporters in the higher income group by 10%, in order to reach 73% (if this figure is at all reachable), this will cost a minimum of one order of magnitude more effort than to increase the support in the lower income group to 38%.

CONTROLLING THE PROCESS

The previously mentioned NASA experience has shown the delicate balance between faster, cheaper and better (or do we have to say "faster and/or cheaper and/or better"?).

One of the previously discussed aspects in this context is the "Speed to Market" aspect, which is often very similar in space activities. Driven by the need for publicly visible performance, and an awareness of the time/cost relation, care has to be taken on how projects are speeded up. Specifically when high reliability has to be reached, as for example in manned space activities, precautions are needed to avoid human mistakes in accelerated programmes. As one example: basically the unmanned Soyuz rocket is technically similar to the manned version, but parts for a manned rocket are always worked on during normal working hours (no shift work) and overtime is strictly avoided. Slater covers this as follows [31] p.262:

> *Speed is not achieved by making everyone work faster though. Speed is achieved by eliminating wasteful steps and procedures in both the development and production processes, by organising more efficiently and by catching mistakes earlier.*

Griffin [32] found out that product development cycle times increased with product complexity and product newness, whereby both elements can be compensated by making use respectively of cross-functional teams and, of a formal product development process. The article provides an overview of different studies in this area, whereby the use of cross-functional teams seems to come out as one of the generally accepted prime contributing factors. Also of interest is the qualitative finding of the study that each major change (with an impact of 10% in the design process) adds an additional 1 to 1.5 months in average to the development cycle.

Empirical studies [33] came to the conclusion that cross-functional teams are more likely to come up with successful products when they are given substantial autonomy – guided by a broad strategic directive – to determine their own specific objectives and results, and especially to develop their own processes and procedures. When it comes to upper management control of product development, less appears to be better. Also these findings go in the direction of creating a common "Knowledge Pool", accessible at all levels by a broader (cross-functional) community.

Based upon a number of interviews with project managers, Sethi [34] analysed the factors, which contribute to the quality of a new product. The findings, summarised in table 3.8, confirm some of the parameters which were mentioned earlier as well as some "common sense" appraisals, namely that quality is influenced positively by the internal quality orientation, good information flow within the company and combined with early feedback from the customer (an additional argument for a PPP approach). The fact that many innovative steps in the process have a negative influence on the quality is equally evident, but the fact that time pressure does not have a negative influence on quality (which was the starting hypothesis) is surprising. However, the author notes the differences between the more conventional and the hi-tech sector; in the latter case, time pressure tends to negatively influence quality.

POSITIVE IMPACT	NEGATIVE IMPACT	NEUTRAL
• Information integration in the team • Customer's input on product development • Quality orientation of the firm	• Innovativeness of new products	• Functional diversity • Time pressure

Table 3.8: Parameters influencing the quality of new products [38]

3.6 SPIN-OFF

Spin-off has been one of the important driving elements towards investing in space activities. Chairing the Special Committee to create NASA, Senator (and later President) Lyndon B. Johnson stated on May 6, 1958 [35]:

> *Space affects all of us and all that we do, in our private lives, in our business, in our education, and in government (...). We shall succeed or fail depending on our success at incorporating the exploration and utilization of space into all aspects of our society and the enrichment of all phases of our life on Earth.*

What makes spin-off so particularly important in space activities is, as mentioned earlier, the fact that only a limited number of products can be used "off-the-shelf" in space programs. However, the contrary is true for products that have evolved from space research and were then transferred to other sectors. Typically this is the case for products, which have to be light-weight and at the same time withstand very harsh environmental conditions (such as large temperature changes).

As can be illustrated in figure 3.6, we can differentiate between tangible and non-tangible spin-off.

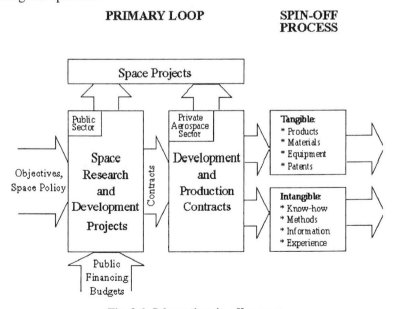

Fig. 3.6: Schematic spin-off process

- In the first loop, objectives by public organisations such as governments or space agencies are given as input, together with the related funding. Agencies define the related projects and programmes, which are executed under contract by the space industry
- Doing so, space industry learns of a number of products and processes; due to its link with the commercial markets, industry marks commercialisation opportunities and produces a number of such "spin-off" products (the tangible flow of the second loop)
- However, there is also an intangible flow: working with rigorous Product Assurance requirements and Project control techniques gives an indirect advantage for industry. Software has to be developed to cope with this and personnel need to be trained. These skills can be used again to improve efficiency in different product categories.

Unfortunately, the general public still thinks that only the famous "teflonpan" is the only product resulting from space activities. It is a good example on how a number of essential, good marketing arguments, are not being sufficiently communicated in support of space activities.

Some illustrations of the spin-off effect are listed hereafter.

- Ensslin [36] made an extensive study on this topic and lists some 50 "spin-off" publications with real applications, coming to a total of some 1400 different applications up till 1987

- NASA [37] publishes a list of spin-off results in yearly reports, grouped in areas as Environmental Management, Health and Medicine, Computer Technology, Public Safety, Industrial Productivity, Transportation and Consumer/Home/Recreation

- NASA estimates that more than 30,000 spin-offs have been successfully implemented for commercial use since 1958 [30] p.81

- ESA is following a similar approach with the so-called TEST (Transferable European Space Technologies) programme. As a summary, some 350 technologies have been presented by ESA over a period of eight years. More than 80 transfer agreements have been concluded with space-technology "donors" through the meditation of "Spacelink" [38]

- One study performed in 1989 has traced 441 cases from NASA's Spin-off magazine. Although 368 cases had acknowledged commercial implementations, only 259 could be quantified. The total benefit from these 259 spin-offs amounts to an impressive 21.6 billion $. [30] p.82

- As another element in this area, ESA [39] has ordered studies, performed by the University of Strasbourg, demonstrating that the spin-off in European industry is very high. Indeed, for each contract placed by ESA, in the period 1977-1991, there was a proven multiplier factor of 3.2 by industry. In other words, for each EURO received via ESA contracts, the company finally obtained 3.2 EURO in overall turnover. It is interesting to note that there are considerable differences found in return depending upon the type of firm, as shown in figure 3.7

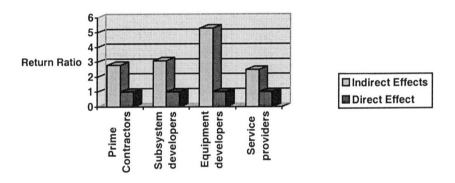

Fig. 3.7 Multiplier factor per type of industry

- A similar study was performed in Canada [40], covering the period 1979-1993. Although the sample of firms was smaller, also here a multiplier factor of 3.5 was found (in this case with a considerable export effect)

- A different approach was followed in a U.S. survey [41], which comes to a multiplier from NASA contracts of overall 2.1 (with a peak value of 5.9 for electronic components). The difference is due mainly to the effect that in this case tangible spin-off was measured in terms of direct jobs created only.

From these examples it is evident that there are plenty of positive studies regularly appearing on this topic, providing sufficient material for debate. Obviously, however, this is not put in a proper marketing context as yet.

From the spin-off effect projects, a considerable number are known but, equally, a large number of them are unknown in relation to their "space-roots". It is far beyond the scope of this work even to try making a non-exhaustive list thereof, nevertheless a number of interesting commercial spin-off products and their space-roots are given in table 3.9

PRODUCT	SPACE ROOT
Fire fighter suits	Apollo suit
Tumor tomography	NASA scanner for testing
Battery powered surgical instruments	Apollo moon programme
Lambda sensor in car catalyst	Apollo Lunar Lander
Anaesthetic gasses monitoring	Apollo suit respiratory system
Non-reflective coating on PC screens	Gemini window coating
Emergency blankets (survival/anti-shock)	Satellite thermal insulation
Mammogram screening	Space Telescope instruments
Heart assist pump	Space Shuttle technology
Plant photon-counting technology	Hubble Space Telescope
Skin cancer detection	ROSAT X-ray detection
Dental orthodontic spring	Space Shape Memory Alloys
Early cancer cell detection	Microwave spectroscopy
Railway scheduling	Ariane check-out software
Coatings for clearer plastics	Material for Shuttle bearings
Fuel cell driven car	Energy source for satellites
Carbon-composite car brakes	Solid rocket motor nozzles
Car assembly robots	Space robotics
Flameproof textile	Ariane protective layers
Lightweight Car frames	Space Shuttle
Fresh water systems	ISS technology
Computer game sticks	Space Shuttle hand controller
Golf shoes with inner liner	Space suit cooling systems
Non-skid road paint	Shuttle booster coating
Corrosion-free coating (statues)	Launch pad protection
Ski-boots flexibility	Space suit design
Health food	Space food
Fuel tank insulation	Ariane polymer blankets
Light allergy protection	Space suits

Table 3.9: Some spin-off examples and their space-roots

MEDICAL SPIN-OFF

A specific spin-off area is the medical one. Probably the main reason is the ethical side of this area and the care taken by medical doctors, from training, not to "raise false hopes". Recent claims that research on the International Space Station may contribute to the knowledge on AIDS, therefore, should be viewed with extreme caution.

However, there is a second spin-off area, which has obvious and direct effects, resulting from the long-duration stay of human beings in space. Here, the approach has to be the complete opposite in order to avoid the generation of a public opinion on "experimenting on humans".

Braak [42] has described the different aspects in a systematic overview, making the differentiation between:

- Life science research
- Medical countermeasures and monitoring of "Man in Space".

Under the first category falls all fundamental phenomenological research, which either has a relation with gravity (like fluid shift), or orientation (like neurovestibular functions). By excluding the one parameter, gravity, over long periods, these phenomena can be studied and one can make a feedback to medical treatment.

An overview of the status of life science experiments can be found in Moore [43], whereby the following categories are considered as prime candidates for research in human physiology:

- Muskuloskelital system
- Cardiovascular function
- Fluid equilibrium and kidney function
- Respiratory function
- Sensory-motor function
- Hormones and metabolism.

The fact that we have astronauts in space has been the source of major emphasis in life sciences. The field has been prepared in Russia by extensive ground tests (like bedrest studies) and, specifically, by launching increasingly complex animals (rats → rabbits → cats → dogs) into space (including the famous Laika-dog). Behaviour in microgravity was filmed with the reassuring fact that animals of increasing complexity could cope better with the effects (e.g. dogs were reported to quieten down fairly rapidly and settle themselves down in space in a fairly relaxed manner). This preparatory process of investigating the effects on increasingly complex biological systems is well described by Yuri Gagarin himself, in [44] pp.218-219.

We know now that exposure to the space environment leads to a number of accelerated biochemical and physiological phenomena, which would take much longer on Earth. Therefore, space medicine has had to develop a number of "countermeasures" to protect the astronaut which, after validation in space, can now directly be implemented on Earth.

It has to be noted that such phenomena evidently are referring to very common and widespread medical problems (specifically those commonly affecting by older people) and, therefore, immediately create a vast amount of interest and potential applications. These are grouped in [42] as:

- **Rehabilitation and motor skills**, i.e. how to readapt after a major impairment event (such as a stroke)

- **Orthostatic Intolerance**, a disease with estimated 500,000 patients in U.S. alone

- **Osteoporosis**, probably the most visible and known direct effect of medical space research nowadays, because approximately 1/3th of women after the menopause suffer from this disease. Studies of the accelerated bone loss can help to develop countermeasures and help millions of people on Earth. In view of such a high number we can even relate this to the cost associated with it. Indeed, osteoporosis is estimated to cost health insurance in Europe some 27 MEURO per day!

- **Physical fitness**, a market which is obviously omnipresent and is also obtaining valuable feedback from astronaut activities

- **Nutrition**, research on astronaut food has been implemented in the "health food" market (estimated at 25 billion EURO yearly) but also found its way into post-operative treatments and other areas, when compact but healthy food needs to be stored or carried (mountain climbing, ocean sailing, expeditions)

- **Biomechanics and backpain treatment**, which occurs as an effect of spine elongation in space and needs to be treated upon return to Earth. Again here, research on the causes and experience with the countermeasures is providing valuable feedback.

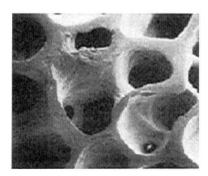 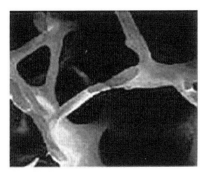

(Photo : ESA)

Fig. 3.8 : Normal (left) and osteoporotic (right) bone structure

One could even extend this list by simply recalling the know effects of spaceflight on the human body, as summarised in table 3.10 [45]. As these effects are constantly monitored and countermeasures are studied, one can safely assume that they will lead in each case to one kind of direct or indirect spin-off for terrestrial applications.

IN-FLIGHT EFFECTS	POST-FLIGHT PROBLEMS
• Space Motion Sickness • Headache / sinus congestion • Bone demineralisation • Muscle atrophy • Orientation problems • Backache • Reduced red blood cell mass • Decreased fluid volume	• Orthostatic intolerance • Reduced exercise capacity • Impaired coordination • Postural instability

Table 3.10: Known effects of spaceflight

Linked to these areas we have to mention the health monitoring of astronauts, which has resulted in considerable progress in the field of

telemedicine. As medical on-board surveillance is largely based on medical teleservices support, telemetry systems had to be developed to bring essential medical data down to earth to the medical experts in order to make a diagnosis.

The same technology can be used to monitor the health of home-patients (saving expensive hospital costs as well as bringing social help for the family) and the development of active expert systems. Non-specialists (for example a stewardess) on board a ship or airplane can measure the essential medical parameters of a patient and transmit them to medical specialists on ground using these systems. Already avoiding the considerable costs associated with emergency landings of commercial carriers (fuel, delays, passenger claims) from having a sick passenger on board has a considerable market potential. Indeed, if this can be reduced to less than half by remote diagnostics (for cases where immediate treatment is not necessary) the savings for each landing are worth several millions of dollars.

Similar technologies can be used in remote areas where medical assistance can be given by general practitioners or paramedics, based upon feedback from telemetry (this procedure is inter alia successfully implemented in inter alia Australia, Canada, Greece (Islands) and Scotland).

This field of applications has been given a lot of attention and therefore it is very useful to illustrate the particularities of medical spin-off. The advantages of telemedicine, for all groups concerned, seem very obvious if we consider table 3.11 [46].

PATIENTS	PRACTITIONERS	HOSPITALS
• Access to healthcare in remote areas • Time/money saving (travel, work) • Easier access to specialised services	• Improved communications • Continuous education/training • No geographical barriers	• Better use of equipment • Quicker and more accurate diagnosis • Better access to patient data

Table 3. 11: Telemedicine advantages

Important progress is reported in specific cases, such as use for military purposes and in remote areas such as Newfoundland, Kenya and Uganda (the SHARE and TETRA projects).

However, the authors [46] point out a number of obstacles, which are significant in part for most medical spin-off applications:

ETHICAL OBSTACLES	MARKET PROBLEMS	LACK OF STANDARDS
• Confidentiality in data access • Integrity (no direct link to the patient) • License system and link to national systems	• Physicians not fully convinced (yet?) • Funding of investments • Dependency on government regulations	• Lack of technical standards (confidential data transfer) • Lack of clinical standards

Table 3. 12: Telemedicine application obstacles

We note that, besides a number of technical elements which are not unusual for a new development, the mainly ethical aspects are considered as potential and major obstacles. Present studies and developments in the field of medical data transfer and encryption, done in ESA [47] may even lead to demonstration to the general public of the confidential treatment of electronic data and demote the "confidentiality" obstacle.

Similarly, efforts to harmonise worldwide medical terminology are ongoing in this context which also would assist in reducing the present drawbacks.

3.7 CONCLUSION

In the framework of a marketing approach, the Space Product plays a very particular role. On the one hand, the technical complexity and indirect contact with the product are leading to the fact that the general public has really problems to identify the product itself. On the other hand, a number of indirect aspects are much more visible to the same general public and therefore require proper attention; in particular, this is the case for launch failures or other spectacular failures (such as the Mars landing events).

Specifically in the case where public funding is used, there is a high sensitivity for such events and this is reflected in the overall attitude towards space activities in general. This, in turn, leads to high emphasis on reliability and quality issues, inter alia by implementing TQM techniques.

A number of elements, which contribute to a successful commercial product are equally applicable to space products. These include the choice of space "products" (missions) which reflect the expectations of the taxpayer and a transparent demonstration of efficiency. This latter aspect, even though adequate tools such as MBO have been available since long, has been ignored in the past and has led to a general feeling of inefficiency from the Space Agencies. In the last decade, important steps have been taken to improve their image in this respect.

The other important factor in this context is the spin-off of space developments, particularly for commercial or other products that are appealing to the general public more directly. The transfer of space research into enhanced medical know-how, to give one example, is one of the major tools that can justify in an objective way the importance of space activities. Similar justification can be found in the large number of products whose existence is directly related to the specific requirements imposed on space products. Unfortunately, only seldom are such products evaluated in a quantitative way in terms of financial return.

Macroeconomic studies indicate a high return factor from space activities, in the order of 2-3 (depending on the evaluation criteria) times the initially invested funds. Again as for space products this is a factor which deserves more appropriate attention.

REFERENCES CHAPTER 3

1. WERTZ, J. and LARSON, W., *Space Mission Analysis and Design* (Kluwer, 3rd Ed., Dordrecht, 1999).
2. LARSON, W. and PRANKE, L., *Human Spaceflight* (McGraw-Hill, New-York, 1999).
3. MESSERSCHMID, E. and BERTRAND, R., *Space Stations* (Springer, Berlin, 1999).
4. MICKLETHWAIT, J. and WOOLRIDGE A., *The Witch Doctors* (Mandarin, London, 1997).
5. ANDREASEN, A., Nonprofits: check your attention to customers. *Harvard Business Review*, (May 1982), pp.105-110.
6. DRUCKER, P., *Managing the Non-profit Organization*, (Harper, New York, 1990).
7. BLOOMQUIST, C., Spacecraft Anomalies and Lifetimes, *IEEE Proc. Ann. Reliability and Maintainability Symposium*, 186 (Jan. 1984)
8. DEMING, W., *Out of the Crisis*, (MIT, Cambridge, 22nd Ed. 1994)
9. OAKLAND, J., *Total Quality Management*, (Butterworth, London, 2nd Ed. 1995).
10. COHEN, S. and BRAND, R., *Total Quality Management in Government*, (Jossey-Bass, San Francisco, 1993).
11. SEBOLT, W., REICHERT, M., HANOWSKI, N. & NOVARA, M., A Review of the Long-Term Options for Space Exploration and Utilisation, *ESA Bulletin*, 101 (ESA, Noordwijk, February 2000) pp. 31-39.
12. PETERS, T. and WATERMAN, R., *In Search of Excellence* (Warner, New York, 1982).
13. COOPER, R., The Dimensions of Industrial New Product Success and Failures, *Journal of Marketing*, Vol. 43(3) (1979) pp. 93-103.
14. VANDEN ABEELE, P. and CHRISTIAENS, I., Strategies of Belgian High-Tech Firms. *Industrial Marketing Management*, 15 (1986), pp. 299-308.
15. McGRATH, M and ROMERI, M., The R&D Effectiveness Index: A Metric for Product Development Performance. *Journal of Product Innovation Management* 11 (1994) pp. 213-220.
16. SOUDER, W. and BETHAY, D., The Risk Pyramid for New Product Development: An Application to Complex Aerospace Hardware, *Journal of Product Innovation Management*, 10 (1993) pp. 181-194.
17. BONNET, R. and MANNO, V., *International Cooperation in Space*. (Harvard University Press, London, 1994).
18. MOOY, S., *Product Communication and Information*, Doctoral dissertation, (TU Delft, Netherlands, 1998).
19. IN 'T VELD, J.: *Analyse van Organisatieproblemen*. (EPN, Houten, 7th ed.1998).
20. BRADY, R.: MBO Goes to Work in the Public Sector, *Harvard Business Review*, 51 (1973), pp. 65-74.
21. GOLDIN, D., *Statement before the U. S. Senate*, (NASA, Washington, April 1998).
22. FERSTER, W.: Study: Faster, Better, Cheaper Method Works. *Space News* (September 6, 1999) p.6.
23. ISBC, *State of the Space Industry 1999* (Space Publications, Bethseda, 1999), p.34.
24. COX, S. and TAIT, N., *Reliability, Safety & Risk Management*, (Butterworth-Heinemann, Oxford, 1991) p.209
25. BERRY, L. and PARASURAMAN, A., *Marketing Services: Competing through Quality*. (The Free Press, New York, 1991).
26. PARASURAMAN, A., ZEITHAML, V. & BERRY, L., SERVQUAL: A Multiple-Item Scale for Measuring Consumer Perceptions of Service Quality, *Journal of Retailing*, Vol. 64(1) (Spring 1988) pp. 12-40.

27. FILE, K., JUDD, B. & PRINCE, R., Perceptions of Quality in the Nonprofit Relationship, *Journal of Nonprofit & Public Sector Marketing,* Vol. 4 (1/2) (1996) pp. 75-87.
28. VANDAMME, R. and LEUNIS, J., Development of a Multiple-item Scale for Measuring Hospital Service Quality. *International Journal of Service Industry Management,* Vol. 4(3) (1993) PP. 30-49.
29. HOFFMAN, K. and BATESON, J., *Essentials of Service marketing*, (Dryden Press, Fort Worth, 1997) p.215.
30. HARDERSEN, P., *The Case for Space*, (ATL Press, Shrewburry, 1997) p.165.
31. SLATER, S., Competing in High-Velocity Markets, *Industrial Marketing Management*, 22, (1993) pp.255-263.
32. GRIFFIN, A., The Effect of Project and Process Characteristics on Product Development Cycle Time, *Journal of Marketing Research*, Vol. 34 (February 1997) pp.24-35.
33. BONNER, J., RUEKERT, R. & WALKER, O., *Management Control of Product Development Projects,* Report 98-120 (MSI, Cambridge, October 1998).
34. SETHI, R., New Product Quality and Product Development Teams, *Journal of Marketing,* Vol. 64 (April 2000), pp. 1-14.
35. MOTT, M., New Markets: The Role of Exploration, in HASKELL, G. and RYCROFT, M., *New Space Markets* (Kluwer, Dordrecht, 1998), pp.85-92.
36. ENSSLIN, K., *Technologische Spin-off-Effekte aus der Raumfahrt*, (Peter Lang Verlag, Frankfurt, 1988).
37. NASA, *Spin-off*, yearly publ. Office of Commercial Programs, (NASA HQ, Washington).
38. ESA, *Impact 2000*, ESA BR-154 (ESA, Noordwijk, 1999)
39. ESA, *The indirect economic effects of the European Space Agency's Programmes*, ESA-BR-63, (ESA, Noordwijk, 1989).
40. COHENDET, P., Economic Rationale for Space Activities, in HOUSTON, A. and RYCROFT, M., *Keys to Space* (McGraw-Hill, Boston, 1999), p.11-11.
41. BEZDEK, R. and WENDLING, R., Sharing out NASA's spoils, Nature, Vol. 355 (9 January 1992).
42. BRAAK, L. Benefits of Space Medicine for Health. *Proc. of the 2nd European Symposium on the Utilisation of ISS,* ESA SP-433, (ESA, Noordwijk, February 1999) pp.491-495.
43. MOORE, D., BIE, P. and OSER, H. *Biological and Medical Research in Space.* (Springer, Berlin, 1996).
44. GAGARIN, Y. and LEBEDEV, V., *Psychology and Space*, (MIR publ., Moscow, 1970).
45. CHURCHILL, S., Introduction to Human Space Life Sciences, in HOUSTON, A. and RYCROFT, M., *Keys to Space* (McGraw-Hill, Boston, 1999), p.18-13.
46. COHENDET, P., VALIGNON, L. & SYLLA, H., Space Assets in the Emerging Telemedicine Market, in HASKELL, G. and RYCROFT, M., *New Space Markets* (Kluwer, Dordrecht, 1998), pp.191-208.
47. DAMANN, V., *private communication,* 15.8.2000

Chapter 4

PRICE OF SPACE PROJECTS

4.1 INTRODUCTION

The Price aspect may at first glance be less applicable in the particular case of public space activities, and therefore will be explained in the nonprofit marketing context initially.

It would be closer to the daily use of terminology to discuss in this chapter the "Cost" of space projects, and in fact a number of techniques relate to this (Cost Control, Life Cycle Cost). The advantage of the "Price" concept is clearly that it reminds us of the fact that we exchange goods; a product is offered in exchange of for example taxes, and this restores the healthy client-customer relationship.

Space projects are considered to be very expensive, although this can be explained on the basis of the very specific environment. However, what is much more critical than the absolute price are the (relative) cost overruns. Of course, in our information age overruns become unavoidably public sooner or later, and will over-proportionally irritate the taxpayer. Sufficient methods are available to at least limit such overruns, such as cost estimating techniques, appropriate contract types, cost control and risk management. They are discussed in this chapter in the framework of overrun avoidance.

Special attention is drawn to the Life Cycle Cost control, because only controlling the development costs, in the past, has often led to budgetary shortages in later phases of space projects.

One cost factor, space insurance, is an inherent part of the risk management approach and therefore will be here treated separately.

4.2 PRICE DEFINITION IN THE NONPROFIT ENVIRONMENT

As mentioned, one may have the impression that the concept of a Price is less applicable for the nonprofit sector. The major reason is that we are traditionally associating terms such as "market" and "customer" with tangible examples from environment. In a nonprofit environment, the target "customers" are not required to make a "sales" decision based upon a cost/benefit trade-off. They are asked to exchange something they value for something beneficial provided by the nonprofit organization.

The costs, or sacrifices requested can be divided into 4 categories:

a) **Economic costs**: this is the case for taxes for government spending
b) **Sacrifices of ideas or views**: examples here are campaigns against racism, whereby people are requested to "sacrifice" their old ideas
c) **Sacrifices in patterns or behaviour**: like the introduction of non-smoking areas
d) **Sacrifices in time and energy**: voluntary services fall in this category.

All nonprofit organizations will request one or more of these sacrifices from their customer. In exchange for this, they will reward them with economic, social or psychological benefits. The related price can take many forms, such as entrance fees, service charges, contributions and so on. Pricing decisions can be based upon full cost or partial cost recuperation and can even at a first sight be zero. In principle one distinguishes following price settings and strategies [1] p. 103:

- **Cost plus**: the costs are calculated and an overhead charge added
- **Competitor matching**: setting the price in accordance with competitors in the market
- **Affordability**: matching what the recipient group can afford
- **Achievement of organisational objectives**: price can be used to achieve the organisation's objectives in terms of the penetration that a given service provision will have.

In addition to this, "Price Discrimination" methods are widely used in the nonprofit sector. The organisation identifies a number of categories of customers and adapts its pricing strategy to their respective abilities to pay. In marketing terms, the different segments are identified and for each segment a different price will be charged. This is for example the case for cultural events, where price reductions are given to students and pensioners, whereas on the other hand price increases are asked to supporting or sponsoring persons, in exchange for some benefits (such as reserved places).

In this way, the nonprofit organisations can adapt its pricing policy. However, this is not the case when costs are covered by tax money. Although there is a modulation built into most taxation systems (progression of taxes), taxes or not "a la Carte". Indeed, in the previously mentioned case, one can decide to participate in the cultural event or not, i.e. upon judgement if the (modulated) price is affordable. Control on the expenditure of taxes is very indirect; it is impossible to refuse part of the taxes if one does not want to support military expenses or, in our case, expenditure for space activities.

Kotler highlights such problem type by analysing the central role of exchange. Two parties exchange one commodity (goods, services...) in exchange for a certain cost (monetary or non-monetary). A favourable exchange is perceived when the ratio of the benefits to the costs is better than for any alternative. Formally, Kotler [2] p.70 assumes that four conditions exist for this process:

1. There are at least two parties
2. Each can offer something that the other perceives to be a benefit or benefits
3. Each is capable of communication and delivery
4. Each is free to accept or reject the offer.

It is evident that in our case this process is not taking place. Whether the individual taxpayer finds space beneficial or not, he cannot refuse to pay for it; moreover he does not have the choice to select alternatives directly.

Under these circumstances it is not surprising that the taxpayer is very sensitive to the price of space projects. This is less the case for the less visible running costs, but the visible parts, such as project cost and, even more so, overruns, are sensitive items. Especially in case of failures, such as the Mars Explorer, the apparent high cost figures gave raise to a public reaction. Such a reaction may not have immediate effects, but in a democratic system pressure will come from politicians on behalf of the taxpayers. This may lead to reduced budgets and even stopping or eliminating complete programmes.

In view of this effect, the apparent costs of space projects and, specifically, the overruns, are paramount elements in any Space Marketing Strategy and will be worked out in more detail hereafter.

4.3   APPARENT COST OF SPACE PROJECTS

There are three major factors leading to an apparent high cost of space projects [3],

a) The near-perfect transparency.
b) The limited possibility to use other developments.
c) The "prototyping" aspect.

TRANSPARENCY

When comparing the costs of a space project with another development project, one soon finds out that there are considerable differences. Other development projects have different sources or better opportunities to "spread" its resources. Examples are:

- Infrastructure is paid from (local) building budgets
- Part of R&D is financed by basic research grants
- National laboratories/universities which participate are financed from different sources
- End-users (for example military) already contribute from the beginning onwards using less transparent budgets.

This is not the case for a public space project. All associated costs are bound to be detailed in the programme declaration and this leads to much higher apparent amounts. Therefore it is difficult to compare such cost with any other government/military project, or even with other similar projects (such as the Shuttle).

USE OF OTHER EQUIPMENT/MATERIALS

The hybrid conditions that all space-segment equipment has to sustain are both atmospheric conditions (even very severe at the launchpad) and conditions in space (microgravity/vacuum/radiation/thermal/solar) and usually makes it impossible to use "off-the-shelf" items. Whereas on the other hand, there is a considerable spin-off from such developments for other sectors and again it makes the "visible" cost aspect very high. An often-quoted example to illustrate this effect occurred in the early Apollo days. It soon became clear that the existing "copying" tools could not handle the considerable quantity of configuration control documents. Therefore, money from the Apollo Moon project had to be invested in developing photocopy-machines. On many occasions afterwards, materials and equipment had to be developed (Mylar, thermal protection tiles,) before the development process could start, adding to the overall development costs.

PROTOTYPING

Often the costs of products resulting from a production series (such as civil and military planes) are incorrectly compared with the costs of a space project

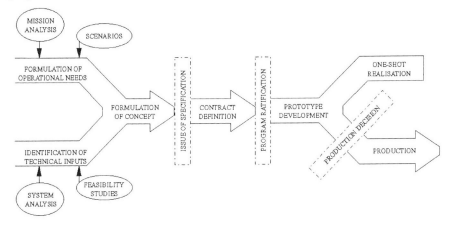

Fig. 4.1 Production flow

However, although both serial products as well as space products have to undergo a similar development process (see figure 4.1), most of the space projects end up as a "one-shot-development" or even as a protoflight model. Therefore, comparing the cost of such unique model with the unit cost of a serial product leads to unfair comparisons. As an illustration from a contemporary project:

The price of a C-17, based upon 210 aircraft, was estimated at some 190 MEURO. Such a figure "sounds" reasonable but if we look into the cost of the development programmes (one prototype and two structural models); we are again back to an order of magnitude of 4 billion EURO (i.e. comparable with a space project like ESA's Hermes project).

However, if we look at the development cost of a number of non-space projects, we note that they all evolve in the same order of magnitude (prices are Cost To Completion, not actualised ones) as for space projects, which is not visible using a unit cost comparison.

PROJECT	ORDER OF MAGNITUDE (MEURO)
Olympic Games Organisation	2,000
VW- Golf	5,000
Concorde	12,000
F 117	7,000
EFA (7 prototypes)	8,000

Table 4.1: Development costs of non-space projects

4.4 THE PROBLEM WITH OVERRUNS

Overruns are a classic problem in public spending. Early examples can be found already in Roman history, as quoted by Gibbon [4]:

The young magistrate (Herod, son of Atticus), observing that the town of Troas was indifferently supplied with water, obtained from the munificence of Hadrian 300 myriads of drachmas for the construction of a new aqueduct. But in the execution of the works the charge amounted to more than double the estimate, and the officers of the revenue began to murmur, till the generous Atticus silenced their complaints by requesting that he might be permitted to take upon himself the whole additional expense.

Later on in history, two effects became more apparent: the "officers of the revenue" stopped murmuring and started shouting, and the Atticuses became more and more scarce...

In the 1950's and in particular during the Korean War, overruns by factors of 200 up to even 800% were noted. Owing to an increased degree of technological complexity and to the need for rapid development prompted by the war, a considerable number of cost-plus contracts were being issued without the possibility for appropriate control. In the winter of 1958, the U.S. Government established a study group at the Harvard Business School to undertake a systematic study in the area of advanced weapons acquisitions. Peck and Scherer [5] studied various development programmes and came to following conclusions:

- For twelve major weapon systems development costs were found (as an average) to be not less than 220% higher than originally estimated, with, in one case, an overrun of 600%.
- For high technology planes and missiles, the average cost overrun was not less than 240%, with a peak of even 1270%.
- For advanced commercial programmes, the average cost overrun was "only" 70%.
- The average development time exceeded the original time estimates with factors of respectively 36 and 40%.

The authors thus conclude [5] p.430:

It is reasonable to conclude from this evidence that, on the average, the organisation charged with conducting U.S. weapons development programmes have been fairly successful in meeting quality predictions, but not in meeting cost and time predictions. This conclusion is no more than common knowledge among government personnel and weapons industry members.

Also similar pressure for success in the space era has led to analogous effects. Holman [6] reports that the total duration of the Mercury project was not less than 2.25 times the originally scheduled one, with a corresponding cost overrun of 120%. On the other hand, the Apollo project is reported to have had a cost overrun of only 25%, which is remarkably good in view of its complexity. This excellent result may be attributed to the fact that probably the schedule had to be adhered to strictly; therefore it also underlines the "faster-cheaper" relation. Under President John F. Kennedy, the Secretary of Defense, Robert McNamara, was given strong instructions to remedy the situation. In his report to the President he stated [7] that a proper use of contract types would effect a reduction of at least 10% in the final costs. This highlights the tendency at that time, when contract types were considered to be an effective tool against overruns. Also at that point in time the first Incentive contracting guides were published and jointly used by DoD and NASA [8] and later implemented in Europe by ESA [9]. However, it has to be mentioned here that the 10% target could not be reached by simply applying different contract types; post-evaluation studies [10], [11] showed that such savings were more in the 3-4% region.

The biggest factor contributing to overruns is undoubtedly the original cost estimate of a project. This is both linked to a technical problem on how to estimate novel projects, and equally, on the wish of the initiator to have a project approved, even by deliberately underestimating the cost. Anthony and Young [12] call this the "Ploys", which can be found in each complex nonprofit organisation. Indeed, the complexity of the organisation is such that these ploys, generally originated at middle-management level, are difficult to detect in an early phase, also because many nonprofit organisations are lacking internal management controls. Examples on how to artificially reduce the initial cost appraisal and still get the approval for a project are adapted from [12] and given in table 4.2.

In the space activities environment we have to mention in this context the Shuttle Transportation System (STS). The U.S. Congress was told in the 1970's that the STS would perform 500 missions in the course of the 1980's, which would lead to a launch cost of 90% less than for expendable rockets, i.e. a launch cost of 10.5 million $ per flight. In reality, some 100 missions were performed in 20 years (a factor 10 difference) and launch costs are in the order of 400 million $ per flight (a factor of 40!).

Common Ploys to obtain budget approvals

'Envelope under the door'
The principle is to match the budget with the allocation, ignoring the expected Cost-to-Completion. Once the (budgetary) envelope is under the door, it will naturally re-expand at the other side. Common tricks are:
- forget maintenance and spare parts
- undervalue utilisation costs
- count on other budgets to cover gaps

'The hidden ball'
Specifically in larger programmes it is easy to hide items which, if presented on their own, would have remote chances to be approved. Examples are:
- new buildings
- new computers
- research programmes

'Keeping up with the Jones'
Under the excuse that specialised staff might become demotivated and go to other establishments, research and lab equipment, PC's and workstations etc. are bought without examining the technical needs. The argument is that the organisation should remain the technology leader (the disadvantage being however that the organisation has to spend resources to debug new software etc.).

'Call it a rose'
Certain requests are typically criticised by governing bodies, such as furniture and buildings. It is therefore easier to get an 'annex to the office' approved than a new building, or 'galley tools' instead of kitchen equipment...

'The Witches and Goblins method'
In 1968, the U.S. Army, as a counterdefense to the Soviet 'Talinin System' requested funds for a new antiballistic missile system. In reality, the latter system did not exist yet and information was based on vague rumours only. The clue of this method is to act on the 'cultural' c.q. emotional circumstances.

Table 4.2: Budget Ploys

In view of potential public criticism, which may endanger considerably long-term budget approvals, it is therefore essential to keep the cost of space projects under control in a consecutive 4 step, interrelated approach:

Step 1: Realistic Cost Estimation
Step 2: Consideration on the Life Cycle Cost
Step 3: An appropriate contractual framework
Step 4: Cost Control and Risk Management during the project phase.

These steps will be elaborated in the next sections.

4.5 COST ESTIMATING

Historically [13], and in analogy with the previously described re-evaluation of contract types, the famous Rand Corporation in Santa Monica was established in the late 1940s by the Department of Defense as a civilian "think tank" to which it could turn for independent analysis. By 1950, Rand developed and improved the most basic tool of the cost estimating discipline, the Cost estimating relationship (CER) and afterwards merged the CER with learning curves. CERs were developed for aircraft cost as a function of variables such as speed, range, altitude and so on and became the basic tool applied acceptably to all phases of aerospace projects.

With the design of large and advanced rocket boosters, around 1957, the CER techniques were transferred to this area as well but were less successful due to the lower number of statistical data. Various refinements took place and one of the best-known ones is the TRANSCOST model of Koelle [14].

Another significant milestone in cost estimation that occurred during the 1970's was the emergence of the PRICE model, from 1975 onwards marketed by RCA. This initiated the era of the parametric cost models. NASA's "Parametric Estimating Handbook" [15] (which is available on-line via the Internet, a considerable example of modern information distribution!) defines this as follows:

> *Parametric estimating is a technique that uses validated relationships between a project's known technical, programmatic, and cost characteristics and known historical resources consumed during the development, manufacture, and/or modification of an*

end item. A number of parametric techniques exist that practitioners can use to estimate costs. These techniques include cost estimating relationships (CERs) and parametric models.

CERs are defined as mathematical expressions or formulae that are used to estimate the cost of an item or activity as a function of one or more relevant independent variables, also known as cost driver(s). Generally, companies use CERs to estimate costs associated with low-dollar items. Typically, estimating these items using conventional techniques is time and cost intensive. For example, companies often use CERs to estimate costs associated with manufacturing support, travel, publications. As they work with costs in analogy to known costs, they are also often called Analogy Models.

One of the most important aspects of a CER is the choice of the right parameter, representative for the previously mentioned "cost driver". In many cases, the weight is chosen, for example for the following cases (all in 1992 K$), X represents the number of kilograms of the first unit costs of unmanned spacecraft [16], pp.726-729:

- Spacecraft bus : CER = $185.X^{0.77}$
- Spin stabilised apogee motor : CER = $58.X^{0.72}$
- Antennas : CER = $20 + 230.X^{0.59}$
- Launch operations = $64 + 1.44.X$

To give another example, for software development, the lines of codes are taken as a parameter, e.g.

- Flight Software : Development = 375 x KLOC
- Ground Software = 190 x KLOC

With, again, the CER in 1992 K$ and KLOC = Thousand of lines of (Ada source) code.

Other parameters are e.g. the impulse needed (for motors) or the aperture diameter (for Infrared payloads).

Parametric models are more complex than CERs because they incorporate many equations, ground rules, assumptions, logic, and variables that describe and define the particular situation being studied and estimated. Parametric models make extensive use of databases by collecting programme technical and cost history data. Parametric models can be developed internally by an organization for unique estimating needs, or they can be obtained commercially. Typically, the databases are proprietary to the contractor or vendor; however, a vendor will most likely share a description of the data in the database in order to build confidence with their users.

Parametric models can be used to discretely estimate certain cost elements (e.g., labour hours for software development, software sizes…), or they can be used to develop estimates for hardware (e.g., satellites, space shuttle spare parts), and/or software systems (e.g., software for air traffic control systems). When implemented correctly and used appropriately, parametric models can be used as a first primary cost estimate.

Parametric techniques have been accepted by Industry and Government organizations for many years, for use in a variety of applications. For example, many organizations have experienced parametricians on their staff who regularly use parametrics to develop independent estimates (e.g., comparative estimates or rough order of magnitude estimates) and life cycle cost estimates (LCCEs). In addition, Industry and Government often use these techniques to perform trade studies such as design-to-cost (DTC) analyses.

In practice, the use of parametric cost models requires a number of prerequisites, namely:

- Data from which the estimate is based (i.e., historical data to the maximum extent possible and of a comparable nature);
- Guidance and controls to ensure a consistent and predictable system operation;
- Procedures to enforce the consistency of system usage between calibration and forward estimating processes; and
- Experienced/trained personnel.

A first experience in Europe with PRICE partially failed because it was based upon U.S. data and cost structures only. Since then European industry and organisations have been developing their own models, such as DASA's TRANSCOST model, ECOS (ESA's Costing Software) and ECOM (ESA's Cost Modelling Software) [17]. The way in which these models are used in the European environment is described in [18]. NASA goes even one step further and provides a number of online cost models [19] and calculators for various projects, such as

- Advanced mission ROM estimates
- Estimates for DSN (Deep Space Network)
- Expendable Launch Vehicles cost model
- Space Operations Cost Model (SOCOM)
- Spacecraft ROM estimates (also for scientific instruments).

An important point, however. is the availability of experienced "parametricians". Parametric models require assessments on such factors as "complexity level", which entails a certain degree of subjectivity. Therefore trained and experienced staff are needed to guide the initiators through the process in order to reach comparable data (essential in trade-off analysis).

Finally, in more advanced stages of the project, "Grass-roots" estimates can be performed (also referred to as "Engineering Modelling"). These require a very detailed knowledge of all parts (also called "Bill of Quantities"), the respective unit prices and accurate estimates of the number of contractors, in terms of equivalent persons per category. Therefore, the accuracy of the inputs required assumes a detailed Work Breakdown Structure and a design which is completely frozen and known. In function of these different requirements, table 4.3 is proposed in [16] suggesting to which particular phase a specific model is most applicable:

ESTIMATING TOOL	PHASE A	PHASE B	PHASE C/D
Parametric model	Primary	Applies	May Apply
Analogy model	Applies	Applies	May Apply
Grassroots	May Apply	Applies	Primary

Table 4.3: Applicability of cost estimating tools

The various tools each have advantages and disadvantages, as illustrated in table 4.4. The authors [20] of this table also point out that cost data need to be compared taking into account a number of adjustments such as:

- Inflation (when using historical figures)
- The learning factor (in case of limited production lines; the learning factor for space projects is often put at 95%)
- The complexity effect (e.g. in case new technologies have to be developed or in the case of time pressure)

- The cultural effect (which can constitute the most important factor; spacecraft manufacturing management culture is estimated to influence the total cost with a factor of two minimum, compared to e.g. missile manufacturing)
- The organisational structure (such as the number of co- and subcontractors).

ESTIMATING TOOL	ADVANTAGES	DISADVANTAGES
Parametric model	• Can be applied at system level • Early identification of cost drivers	• No traceability to historical data
Analogy model	• Early quick look • Traceability to historical data	• Sensitivity to programmatic aspects
Grassroots	• Proves WBS element costs • Uses available historical data	• Time-consuming and expensive • Requires detailed design description

Table 4.4: Comparison of estimating models

Generally speaking, Organisations have access to a variety of techniques to develop reliable cost estimates for new projects. Such techniques should be capable of estimating a cost within a 20% error margin, if properly used. Therefore there is no a posteriori excuse for excessive cost overruns from this point of view.

4.6 DESIGN TO LIFE CYCLE COST

The term Design to Cost (DTC) has gained a lot of popularity in the 1980's and 1990's. However, there is very little proof of widespread practical implementations of this technique, specifically in public space programmes. The main reason is probably that most design is based upon performance. In an exceptional number of cases, schedule is an important factor and will drive the programme. (As an example: if the Giotto satellite had not been launched on 2 July 1985, it would not have encountered Halley's Comet on 14 March 1986 and would have been stored for some 76 years, until the Comet passed close to the Earth again...). Due to the nature of technically driven organisations, cost is the third, most important parameter most of the time; very few programmes are known to have had cost as the prime design driver.

However, there is another more fundamental problem. Most of the time, in budget driven environments, Design to Cost is interpreted as "Design to Budget". On the basis of yearly budget allocations, managers will design the payment profiles to fit the budget, which irrevocably leads to increased Cost to Completion as most of the time it will result in stretching the programme. In rare cases, external pressure will even ask for a de-scoping of the project. However, the range for manoeuvre is very limited if it exists at all. The further on the programme is advanced, the higher the percentage of costs is committed. Furthermore, with hardware costs only being a fraction of the total costs in space projects (varying between 20-40 % only, respectively for manned and unmanned programmes), it is evident that even a saving of 5 % overall is difficult to reach from the hardware part only.

All these problems considerably influence the Life Cycle Cost (LCC) in most projects during the development phase. The project manager (who is most of the time not responsible for the operational phase and costs anyway) will concentrate on reliability requirements, seldom on operability and maintainability. Hodge [21] points out that presently the operational costs in the NASA budget are approaching the 50 % mark, compared to 20 % some 20 years ago and, therefore, vigorously defend concentration on Design rather than Life Cycle Cost.

Industry is more used to such approaches, design of for example cars is mainly driven by the production and maintenance aspects. Similarly in aircraft manufacturing, the design is driven by such parameters as MTBF (Mean Time Between Failures) and MTBR (Mean Time Between Repairs) which are the major design parameters. These can be only met if the necessary steps are built in during the early design phase (such as Line Replaceable Units, LRU's) to achieve quick repairs and overhauls. End-to-end developers will first examine the operational cost, as they know, with an expected market prognosis, how many people will be willing to pay. Therefore they need an indication of the operational cost per unit (example: a car navigation system); from that starting point they can calculate backwards and define such design parameters such as reliability and accuracy.

The problem in the public sector is that low-cost operations usually demand higher development costs, which in turn means larger front-end programme costs. Therefore, an operational concept must be developed early enough in the programme to have an effect on the design process.

Awareness of the importance of operational costs in space projects is surprisingly low. On the contrary, for example for weapons systems, this is a more common practice and such knowledge using a number of rules of thumb, typically like the ratio 1:3:6 for design:development:operations for specific weapons systems is frequently used. Such figures have been studied in depth by the U.S. DoD and form the basis of strict System Engineering guidelines, (see also [16] p.694 and following).

Moreover, there is a considerable difference between the Life Cycle Cost spent and committed. Studies on complex systems have demonstrated that already after phase A, when only a fraction of the budget is spent, approximately 70% of the Life Cycle Cost are fixed! This considerable difference can be illustrated as per figure 4.2.

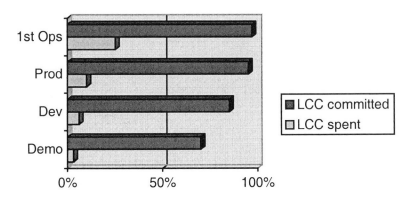

Fig. 4.2: Commitment of Life Cycle Cost per phase

Therefore, it is evident that industrial developers carefully examine each design step in the first phase of the project, an requirement that is even more important in the case of serial production (cfr. Communication satellite systems). We can group the methods for Life Cycle Cost reduction in three categories:

1. Non-value added activity elimination
Whenever a point is reached whereby reduction is needed, the use of so-called "Value Charts" [22] can assist in reducing the costs, without significantly reducing the quality of the final product.

2. Trade-offs

This technique is only valid in an early project phase. There are generally many options that will achieve the objectives. The technique consists of calculating the LCC of each of the options before going into the prototyping phase.

3. Activity-Based Costing

The ABC (Activity-Based Costing) technique originated in production economics. Recently, however, it has also been demonstrated that it could be used in non-commercial and service markets. A full description is given in [23].

Goldin [24] presented as a major achievement the reduced Shuttle cost. His keyfigures are presented in table 4.5

PARAMETER	1993	1997
Operations Cost	2.9 billion $	2.5 billion $
Av. Cost per flight	600million $	400 million $
Contractor Workforce	21,440	16,510
Overtime percentage	6.1 %	4.2 %

Table 4.5: Shuttle operations cost reductions

Although cost control mechanisms are described extensively in [25], one can safely assume that the major reduction effect is the "outsourcing" of the Shuttle operations to industry, leaving to them the task of controlling and reducing the cost. The degree of outsourcing depends on the nature of the activities under consideration and the control the public authority wants to keep, as schematically shown in table 4.6, adapted from [26]:

FACTOR	PRIME CONTRACTOR	PRIVATISATION	COMMERCIALISATION
Management	Public/Contractor	Contractor	Contractor
Daily operations	Public entity	Contractor	Contractor
Manifest control	Public entity	Public/Contractor	Contractor
Asset Ownership	Public entity	Public entity	Contractor
Financial Liability	Public entity	Public entity	Contractor

Table 4.6: Responsibility matrix in outsourcing

Table 4.6 clearly shows that the financial risk can be put only in the hands of the contractor once every form of control or involvement by a public entity is fully abandoned. In cases where for example public or non-commercial aspects are concerned (such as public transport and health care), there is a risk that under full commercialisation public authorities have no direct influence anymore in case of a conflict of interest (e.g. reduced prices for pensioners or free medical care).

Ignoring the Life Cycle Cost has led in the past to a number of dramatic events, such as putting perfectly working satellites into dormant mode when the operational budget was exhausted, or flying fully reusable hardware only once, when no operational budget for reflight was available. A similar effect can be noted on maintainability aspects. Admittedly, the Russian designers of the first Sputnik certainly did not have to take into account repair or maintenance possibilities, but at a certain point in time system designers should have done this. Probably this was not forgotten, but sacrificed in favour of lower development costs. This has led to a number of situations where the taxpayer has witnessed the effects with disbelief. Bromley [27] mentions the considerable loss of image for NASA when it became evident that the Hubble telescope was not designed for easy repairs and at periods in time when the Shuttle fleet has been grounded for long periods due to maintenance problems. Augustine notes the same effect in military aircraft maintenance. Extrapolating budgets and aircraft costs he concludes in one of his famous "Laws" [28] p.111,

> *In the year 2054, the entire Defense budget will purchase just one aircraft. This aircraft will have to be shared by the Air Force and Navy 3,5 days per week except for leap years, when it will be made available to the Marines for the extra day.*

4.7 CONTRACTUAL PROVISIONS

Once a fairly objective cost is estimated, there are techniques to make a relationship between this initial estimate and the final price, by choosing the appropriate contract model. The principles of this technique have been brought forward in [29] and can be described as follows.

Technological development contracts have proven that the traditional types of contract were an inadequate means of controlling the expenditure and management. In order to remedy this, a number of "intermediate" types of contract have come into being, most of them containing "incentive provisions".

Traditionally, two types of contracts were used: Firm-Fixed-Price and Cost-Plus-Percentage-Fee contracts.

Under Firm-Fixed-Price (FFP) conditions, the most important element is the precalculation. Based upon this precalculation, both parties agree upon a price, which is relatively independent of the "a posteriori" actual costs. It is evident that, in this case, the contractor takes a considerable risk whereas the client knows exactly how much he will have to pay finally (provided of course, that his specifications were sufficiently "sound" and that he does not introduce extra requirements afterwards. If the client does not respect these principles, the contractor will "claim" extra costs).

On the contrary, under Cost-Plus-Percentage-Fee (CPPF) conditions, the postcalculation plays a paramount role. The precalculation will only be indicative as a (non-binding) target. The contractor is entitled to charge all justified costs (and, as such, bears a minimum financial risk), whereas now the client will only know the final costs at completion of the contract. Mostly overruns become evident only towards the end of a project and, in the past, they have been considerable as a result.

Mainly large development projects made it evident that these traditional types of contracts could not cope with the actual circumstances. Consequently, such large development projects were transferred more and more from the industrial into the governmental and even supra-national sectors. These sectors are, in general, less "risk-taking" minded in view of their source of financial (i.e. public) support. On the other hand, it is almost impossible to find a contractor accepting the risk for such high-technological and long-term projects and willing to bid under fixed price conditions.

It is mainly for these reasons that governments have come to experiment with a number of "intermediate" contract types (i.e. intermediate between the previously mentioned extremes, FFP and CPPF). A common feature of these intermediate types is the linkage between pre- and post-calculation. In fact, "targets" are fixed in the precalculation phase and reference is made to these targets at the final profit or price determination. As the main purpose of such techniques is to give the contractor an incentive to stay within the targets, these types of contracts are also often referred to as "incentive" contracts. A wide variety of such contract types has been developed during the last 50 years, although some of the variants have been abandoned in the meantime. The contract types, which are used most frequently nowadays, are briefly described hereafter

DIFFERENT TYPES OF CONTRACTS

The two extreme types mentioned earlier belong to two groups of contracts:

(1) Cost-Reimbursement Contracts: In cost-reimbursement contracts the client is required to reimburse all allowable, allocatable and reasonable costs that the contractor can prove he has made.

(2) Fixed-Price Contracts: In this case, the contractor has the obligation to deliver a final product for a specified price, as contractually agreed. (Note that, particularly in the construction industry, these types are also called "lump-sum contracts").

COST-REIMBURSEMENT CONTRACTS

CPPF: Cost-Plus-Percentage-Fee Contract
This is definitely the most straightforward type of contract, from a contractor's point of view. All justified costs are paid and the fee is added as a fixed percentage. The target cost is the estimated cost to completion, whereas the target fee is the fee payable if the actual cost equals the target cost.

CPFF: Cost-Plus-Fixed-Fee
In this case, costs are reimbursed but the fee remains constant, whatever the actual costs.

CPIF: Cost-Plus-Incentive-Fee
CPIF is similar to CPFF, but in this case the fee may vary up or down within set limits and in accordance with a formula tied to allowable actual costs.

We further identify the following elements:

Sharing Formula: The expression (normally in percentage terms) of the basis of the client and contractor's cost-sharing arrangements;

RIE, Range of Incentive Effectiveness: In general, the fee has an upper (maximum fee) and a lower (minimum fee) limit. The band of costs over which the incentive provision is mainly operative is called the RIE.

FIXED-PRICE CONTRACTS

FFP: Firm-Fixed-Price Contract
This is a one-price contract and the price is not subject to any adjustment unless there is a change in the scope of work required under its terms. The profit in this case can even become negative.

FPI: Fixed-Price-Incentive Contract
In this type of contract, a target fee is fixed, but the fee will be determined when actual costs are known. In this case, a ceiling price is determined, which is the maximum price accepted by the client for fee determination: for costs higher or equal to this ceiling price the profit = 0. In general, also, a maximum fee is determined. The difference between this type of contract and the CPIF contract is that, in this case, the upper limit is fixed (ceiling price); in the FPI case the contractor can suffer a considerable loss if his costs exceed the fixed ceiling prices. In the case of CPIF, the worst case for the contractor is no profit; his costs will be reimbursed anyway.

Note that we can represent some of the contract types mentioned by the following relation:

P = BF + s(TC-AC), where:

P= Profit
BF= Basic Fee
s= sharing ratio
TC= Target Cost
AC= Actual Cost (as accepted by the client)

Then:
For s = 0, we have a CPFF contract;
For s = 1, we have a FFP contract;
For 0 < s < 1, we have a CPIF contract.

SOME VARIANTS

So far we have only considered cost incentives in our presentations. However, we should often consider other incentives, such as:

Performance Incentives: A positive or negative fee is paid upon obtaining a certain technical performance. Imagine a satellite payload with a target mass of 50 kg; a performance incentive on mass reduction could be included.

Delivery Incentives: In cases where delivery time is important (e.g. payload for a satellite with a predetermined launch date), (stepwise) incentives are often included.

It should be noted here that these incentives, certainly in the case of cost-reimbursement contracts, should preferably be linked with cost incentives (otherwise, the contractor could incur important costs in order to reach the maximum incentive fee). Consequently, we often find these as multiple incentives; i.e. a combination of cost incentive with performance and/or delivery incentives. In this latter case, the total fee is described as:

$P = BF + s(TC-AC) + DI + PI$ where:

P= Profit
BF= Basic Fee
s= sharing ratio
TC= Target Cost
AC= Actual Cost
DI= Delivery Incentive
PI= Performance Incentive

AF: Award Fees

At the beginning of a contract, an amount is determined for awards, the Award Pool. This award pool is divided into evaluation events, and (subjective) measurement systems are developed and mutually agreed prior to the start of performance (e.g. rating system). At each event, the contractor can earn (part of) the attributed award. In general, a (low) base fee in the order of 2-3% is added, independent of the performance.

The most common form found in practice is a CPAF, Cost-Plus Award-Fee contract. However, award fees can be added to any type of contract.

SYSTEMATIC CONSIDERATIONS ON INCENTIVES

In one study [30], a systems approach was applied to the main incentive parameters. This approach was largely based upon the systems work of in 't Veld [31] and consisted of the development of a qualitative model for each parameter and afterwards, whenever possible, the quantification of this model. Quantification was merely based upon the use of data collected within the ESA pilot environment. Important results are noted here, for more detailed analysis and the basis of formulae, reference is made to this study.

Cost Incentives

The most important aspect when applying cost incentives is the sharing factor. A relation can be found between this sharing factor and the general and administrative expenses, as follows:

s = A / (1+A)   where:

s= sharing ratio
A= Overheads, expressed as a part of the direct costs.

For important development projects involving high-technological support, this formula suggests that sharing should be in the order of 35% (65/35).

Delivery Incentives

Overtime payments to the contractor's staff are considered to be a key element when this incentive is determined. A formula has been developed which gives the following relation:

d = m∗C∗($\frac{R}{0.83}$ - 1) / T

Where:
d= delivery incentive per time unit;
C= project cost;
m= manpower-related part of the project cost;
R= applicable overtime ratio (generally 1.5);
T= contractual delivery time.

This results, again for high-technological projects in typical delivery incentives in the order of 1% per month of total contract costs. Different considerations were applied to penalties, leading to typical penalties (in the same environment) of 0.7% per month.

Appropriate choice of a contract type

There is no doubt that the wide variety of contract types and incentives will make the inexperienced user reluctant to enter into this unknown field. Furthermore, it is evident that no unambiguous solution can be found for each specific project; each contract type should be tailored to the specific requirements of the project. Both parties, the contractor and the client, will meet at the negotiation table. It is evident that, for the same project, there will be differing views initially concerning the (preferred) contract type. Unfortunately, most interests are highly conflicting: an advantage for one party represents an equivalent disadvantage for the other party. This leads us to our first conclusion: the contract type chosen is, in general, the result of a compromise, acceptable to both parties. The following six criteria were chosen to deduce the appropriate choice:

(1) **Cost Uncertainty**: cost and associated risk will play an important role for both parties. However, associated elements such as guarantees, reputation, payment currency etc. will also be taken into consideration.
(2) **Technical Uncertainty**: if a contractor feels that the technological background needed falls outside his field of competence, his cost uncertainty will be influenced. New technologies are inherently linked to uncertainties and risks.
(3) **Available Extra Resources**: in the case of overcapacity, the contractor will be more flexible; cost overruns will be less painful due to coverage of overheads.
(4) **Schedule Criticality**: on the whole the contractor will have a good feeling if the schedule, in general imposed by the client, is a feasible one. This will again influence his decision, particularly when delivery incentives are involved.
(5) **Performance Criticality**: the rationale set out in the previous criterion also applies to performance incentives.
(6) **Long-term Motives**: the contractor may have a number of motives which will make him act more flexibly during the negotiations, such as continuity, spin-off, chance for follow-up contracts and so on.

In order to present results in a systematic way, a decision network format has been chosen. The resulting decision trees are represented in [30] p. 123 and make it possible to select the most appropriate type of contract for each specific case, on the basis of systematic and objective considerations. Again this methodology should not be considered in isolation and must be handled as an integral part of a cost management programme. If the initial cost has been incorrectly set, the contractual conditions cannot remedy this.

4.8 COST CONTROL AND RISK MANAGEMENT

Cost overruns do not appear suddenly, and very often they are preceded by schedule delays. Therefore cost control shall take a proactive approach instead of a posteriori accounting. Also here, the necessary techniques are well known and proven. If we simply refer to the S-shaped cost versus project time curve, any manager knows that this is a given fact for each and any project. Still, whenever such deviations from this standard shape become noticeable, it is amazing to note how often the consequences are ignored.

Within the space community, the phased approach, introduced by NASA, was designed to overcome this problem. The underlying idea was to evaluate the situation at each critical point in time, both from a technical, schedule and cost point of view, and to decide if the next phase would be started. Due to the well-defined scope of each of the phases, it also gives a possibility to transfer the project from one contractor to another one in case of inadequate performance. As the commitment to the programme increases rapidly with each discrete stage, this feedback mechanism is only effective during the early phases. The different phases are often named differently depending on the Agency or the organisation. The same differences are valid for the responsibility assignment. For Europe, the generally accepted distribution is represented in table 4.7 [32] p.5.7.

PHASE	INPUT	OUTPUT	MAIN ACTORS
O / A	Mission Statement	Mission Architecture	User + Implementation Manager
B	Mission Architecture	Detailed Design	Implementation Manager + Industrial Builder
C/D	Detailed Design	Hard- / Software and Launch	Industrial Builder + Implementation Manager
E	Space and Ground Segment	Data / Service	User

Table 4.7: Phased Space Mission approach

After phase O/A (the feasibility phase), only a small portion of the total programme cost is spent, typically less than 1%. Evidently, it is not too difficult to stop a project at such stage. Often two contractors are put in competition during such phase.

Phase B (the programme definition phase) is more critical. The programme cost, schedule, performance and the most likely technical solution will have been established in the course of this phase. It is in this phase that cost should be continuously assessed as a function of the variety of options; it is also very important that the Life Cycle Costs are envisaged during this phase. Stopping the programme at this phase will lead to considerable costs, even more if Long Lead Items have been ordered. The danger point is here if the initial cost estimate deviates considerably from the revised one. Such a situation occurred in the case of the European Hermes project, which was terminated after phase B. For a "classical" satellite, this phase represents some 10 % of the programme costs, but at the end of this phase, some 80 % of the cost elements will be determined and fixed.

In general phases C/D are combined and cover detailed design, development, construction, test and evaluation, culminating in the actual launching of the various elements of the space segment. In this phase the role of cost control shifts rather towards sound definition of end items and, above all a tight control of changes. Design has to be frozen at an early stage, even if new ideas continue to appear afterwards (the "cream on the cake" for contractors). In classical space projects, the costs of this phase are approximately one third of the total programme cost.

Phase E depends largely on the type and lifetime of the system. Whereas this phase was some 20 % of the costs in the past, this phase tends to absorb more costs in view of longer operational lifetimes of the space systems (we should not forget that some of the earlier satellites, specifically in low orbits, only had lifetimes in the order of weeks, even days). This increase is leading to an equivalent increase in the emphasis on Life Cycle Cost.

We have to note here that cost control in the past encountered more difficulties due to the fact that for example standards of reporting were different in European Organisations (also to the big disadvantage of European industry, which had to maintain in parallel different systems and software as a consequence). One major step forward in this field was the creation of the European Cooperation for Space Standardisation (ECSS) [33] which developed and still refines, with ESA taking the role of secretariat, a number of commonly accepted standards applicable for all European space entities, public as well as private. Also for Cost and Schedule management such standards are now generally accepted and implemented [34]. A proven method is to require contractually a number of interrelated documents and reports in a well-defined way, covering elements such as

- Definition of project and status (project control plan, management plan)
- Project structure (Product Tree, Work Breakdown Structure, Work Package Description)
- Planning and schedule (Schedule Tree, Networks, Milestone trend chart)
- Financial and change control (Estimation-at-completion, Contract Status Report, Geographical Distribution Forecast, Change Status List).

Specifically if this information has to be delivered following a defined Management Information System, i.e. by electronic means, a transparent structure which will indicate early warnings can only result from this. Eventually such transparency has proven to be in the interest of all parties concerned, in terms of feedback and remedial actions.

RISK ASSESSMENT

Risk assessment is a more active tool in this respect; it is designed to avoid, anticipate, mitigate and control risks and to allocate optimally programme/project resources where they are needed to ensure success of the programme. From this point of view, the possible impact of any change can be assessed by its influence on the three main project parameters, cost, schedule and performance.

Based upon the evaluation and the expected impact, management can then take a decision at programme level, such as adding more money to one element (avoiding a multiplication of cost in other elements), extending the schedule (avoiding claims for stand-by costs or reducing the performance requirements). The advantage over the previously described phased approach is that risk assessment is a continuous and iterative process. The frequency of the assessments can be modulated (typically higher during the first phases of the project when uncertainty is equally higher). However, above all it replaces reactive management with proactive management. A typical risk assessment follows four steps, summarised in figure 4.3 as follows:

Step 1: **Definition of the Risk Management Policy**

This step is performed at the beginning of the programme and identifies inter alia the goals, the levels of risk evaluated and reported and the frequency of such assessments. Of importance is a consensus with the project managers at this stage, in order to have ensured cooperation.

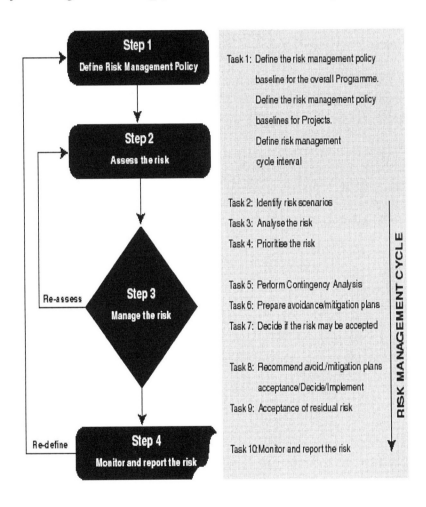

(Courtesy: Belingheri, Tosselini /ESA)

Fig. 4.3: Risk Management Flow

Step 2: **Risk Assessment**

Risk identification is done on the basis of a set of structured questions, which are posed to individual experts using interviews. Experts are nominated from inside and outside the project team, with the consensus of the project manager. The utilisation of experts who are asked a set of questions is a well-known methodology based on the Delphi method. On the basis of the identified risks, an analysis is made whereby the risk magnitude is measured as:

$$R(n) = P(n) * I(n)$$ where:

R(n) = risk of scenario n
P(n) = probability of occurrence of scenario n
I(n) = likely cost/schedule impact of scenario n

The individual risks are then plotted in a probability/impact grid as per figure 4.4, where three areas are defined:

Avoidance area: risk not acceptable
Mitigation area: risk acceptable after optimisation
Acceptance area: risk negligible

In a next step, risks are ranked in order of priority.

Step 3: **Risk Management**

After a contingency analysis what risks can or have to be accepted are defined and for which risk avoidance/mitigation plans will be prepared. Management has to decide in this stage either to accept and implement such an avoidance/mitigation plan or to accept the residual risks.

Step 4: **Risk Monitoring**

Basically, this covers the systematic control and tracking of the implementation plans selected in the previous steps. A report is produced and the risk is tracked during the life cycle of the project.

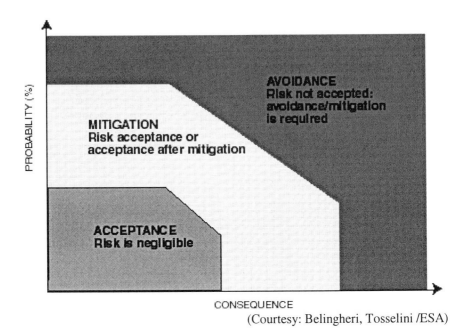

(Courtesy: Belingheri, Tosselini /ESA)

Fig. 4.4 Risk Impact versus Consequence diagram

As a practical example, this procedure is now followed in ESA for the ISS development whereby in the development phase a new risk assessment is performed twice a year, with evaluations at project management level each quarter. Also for Risk Assessment, an ECSS standard is being issued [35]. The technical oriented side of ESA's Risk Management approach is described in [36].

Another Risk management approach, worth mentioning here, is the "Risk Pyramid" approach, as it has been applied to complex Aerospace Hardware. In this approach, which is more geared towards industrial chance of success (but which also could for example be used to evaluate PPP alternatives), three categories of risk are considered, namely:

- Technical risks (potential problems arising from technological problems, design or manufacturing problems)
- Market risks (risks of changing market orientations or competitive actions)
- Business risks (risks resulting from political or economic factors).

The total risk is then calculated according to a scoring formula, namely

$$R_i = (\sum W_{ij} s_{ij}) / R_I$$

Whereby
R_i = the ith risk dimension
W_{ij} = the importance weight of the jth risk source for the ith dimension
s_{ij} = the risk consequence for ij
R_I = maximum value possible.

In [37], the approach is applied to the SPACEHAB module concept and came to the following conclusions:

- The technical risk was assessed at 30%, hence relatively low
- The business risk was 54%
- The market risk was 60%.

For the two latter aspects, a number of corrective actions were suggested. The article was written in 1993 and the assessments can now be validated. Technically the SPACEHAB concept worked out very well, but problems with inter alia flight opportunities have indeed considerably influenced the market position.

4.9 SPACE INSURANCE

Space insurance has been a considerable cost factor and deserves to be covered separately under the "Price" aspect.

Initially, space projects were "self-insured", as is the case for most government property. The situation changed with the arrival of commercial projects, and the first commercial satellite, Early Bird, was covered for prelaunch risks for a sum of 3.5 million $. In 1968 the first "launch risk" insurance was concluded, and in 1975 the first "in-orbit performance" insurance contract [38]. The first ESA satellite insured was OTS, also because the interrelation with the commercially important MARECS programme.

We can distinguish between [39]:

- **Pre-launch and Construction Insurance**

This risk is usually assumed by the manufacturer and covers the damages, which may occur during assembly, integration and testing; during the transport; and associated to launch site activities (fuelling, integration in launcher).

- **Launch and Commissioning Insurance**

This provides coverage from lift-off till commissioning, often in the order of 180 days, when the satellite has reached its final orbit and is checked out and accepted by the client. Therefore the risk can be the loss of life of the satellite or a loss of operational capacity.

- **In Orbit performance insurance**

This covers the proper functioning of the satellite during its operational lifetime, often in renewable yearly phases. Historically, most losses were associated with the launch phase, as can be noted from figure 4.5.

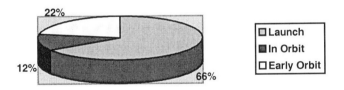

Fig. 4.5: Satellite losses per phase (status 1999) [38]

Therefore, it is evident that many users, such as ESA, concentrate the insurance efforts on the launch phase.

Due to the evolution in launcher reliability and associated losses, premiums have fluctuated considerably. In the early 1980's, premiums were in the order of 7% for a Shuttle launch and 8 % for other launchers. Considerable satellite losses in the middle of the 1980's, as illustrated in table 4.8 [40], led to premiums as high as 25%.

YEAR	SATELLITE LOSS	LOSS (IN MILLION $)
1979	Satcom 3	77
1982	Insat 1A	70
1984	Westar VI / Palapa B2	180
1984	Intelsat V F9	102
1985	Leasat F4 (repaired in orbit)	84
1985	ECS 3	65
1986	Intelsat F14	82
1988	G Star III	60
1990	Superbird A	170

Table 4.8 Some important satellite losses in mid 1980's

Improved launcher performance has led to a gradual reduction of the premiums in the 1990's (7 to 15%) and a considerable overcapacity on the insurer market. The combined capacity of space insurance underwriters was more than 1.2 billion $ during the years 1998 and 1999. Some underwriters have left the space business since then, but the 2000 capacity is still reported to be in the order of 1 billion $ [41].

This, in its turn, has resulted in a proposal by the insurers in Europe to insure all ESA launches at a rate of 4.5% of the project cost (provided certified launchers are used) excluding third-party liability, in-orbit insurance and launcher qualification flights. This offer is valid, provided a long-term contract (e.g. 5 years) is made in order to spread the risks in an acceptable way [42]. It should be noted also thát the insurers, internally, are counting on a gradual risk reduction due to higher launcher reliability, such as, for the case of Ariane5,

- Ariane 505-529 (5th flight till 29th): appr. 4.5%
- Ariane 530-554: appr. 4 %
- Ariane 555 onwards: appr. 3.7%.

Evidently, insurers request a clear definition of risks, and therefore a number of technical parameters will be fixed for each individual case. Therefore, besides the normal contractual and legal conditions, an insurance contract will have to cover such items as:

- Launch delay consequences
- Satellite performance specification (such as positioning tolerances)
- Definition of "Total Loss"

- Partial Loss factors
- Proof of Loss
- Salvage / Post-loss Adjustments
- Abandonment
- Use of Spacecraft after Loss Payment.

An element that will gradually play a role in this context is in-orbit repairs. Whereas in the past few such activities were foreseen, this factor, after the spectacular Hubble Telescope repair mission (see figure 4.6), will result in increased reparability requirements for future satellites as a design parameter.

Fig. 4.6: Repair in orbit of the Hubble Space Telescope (photo: NASA)

4.10 CONCLUSION

The concept of price has a different meaning, depending upon the specific nature of the nonprofit organisation. Nevertheless, every nonprofit organisation has to be aware that a certain "sacrifice" is asked of their "customers", even if it is a non-monetary one.

In the public space environment the matter is complicated because financing runs over non-voluntary contributions in the form of tax money. Even if research has proven that the willingness to pay more (the price-elasticity) is much higher in the case of public goods than in the case of consumer goods [43], the taxpayers have only an indirect hold on this process and this increases their critical attitude.

However, it shall never be forgotten that in a democratic system with higher transparency the reaction time of such an indirect effect is getting shorter. Important space projects such as the International Space Station have been "marketed" to the general public also on the basis of the political dimension. It is evident that the opposing political party will be extremely vigilant to point out any deficiencies under these circumstances that, ultimately, will be reflected in cuts in public space funding.

Two aspects, which will always mobilise the taxpayer, are the cost of space projects and cost overruns. For this reason they deserve special attention from a Marketing point of view, besides their obvious economic importance.

After more than 40 years of space management experience, sufficiently mature methods are available to control both costs and overruns. There are good methods available to do reliable cost estimation and adequate contractual tools to link the "a priori cost estimate" to the "a posteriori payment". In the course of a project, cost control and especially risk management tools can be used to control the overruns.

With the longer operational lifetimes of present space systems, the Life Cycle Cost is becoming of growing importance. Bringing in early experience from the private sector during the design phase is probably the best solution for this; the previously described PPP concept could present an excellent modus operandi in this respect.

There remains the "apparent" high and partially unavoidable cost of the innovative environment where only prototypes are developed. This cost can be explained in objective terms to the general public using the right arguments. Here lies one of the most important tasks of Space Marketing.

One of the arguments that can be used is to express the cost in relative terms:

- DARA distributed in Germany in collaboration with some important newspapers, a leaflet with themes as "What is Space" and "Why Space Activities". The presentation to the general public that space in 1993 was costing each German only 22 DM (appr. 11 EURO) per year or 6 Pfennig (appr. 3 Eurocent) per day was in general received with astonishment
- NASA calculated during the Apollo times that the expenditure per capita for the programme was still less than the expenditure for … popcorn
- If you tell to the beer-loving Belgian that manned space activities only cost him one pint of beer per month, this will certainly have more effect than academic figures expressed in GNP
- Public spending on weapons and armament is one magnitude higher worldwide than spending on space activities
- On average every citizen living in an ESA Member State spends 7.3 EURO on space (in 2000), about as much as a cinema ticket.

REFERENCES CHAPTER 4

1. SARGEANT, A., *Marketing Management for Nonprofit Organisations*. (Oxford University Press, Oxford, 1999).
2. KOTLER, P. and ANDREASEN A., *Strategic marketing for Nonprofit Organizations*. (Prentice-Hall, Englewood Cliffs, 3rd ed. 1987).
3. PEETERS, W., *The appropriate use of Contract Types in Development Contracts*. ESA STR-222 (ESA, Noordwijk, 1987).
4. GIBBON, E., *The Decline and Fall of the Roman Empire*, (Vol.1, 1962) p.28
5. PECK, M. AND SCHERER F., *The Weapons Acquisition Process, an Economic Analysis,* (Harvard University Press, Boston, 1962)
6. HOLMAN, M., *The Political Economy of the Space Program*, (Pacific Book publ., California, 1974)
7. MCNAMARA, J., *First Armed Services Progress Report to President Kennedy*, (8 July, 1963)
8. DOD and NASA, *Incentive Contracting Guide*, NASA Ref. 5104-3A (U.S. Government Printing Office, Washington, 1969)
9. ESA, *General Clauses and Conditions for ESA Contracts*, ESA/C/290 Rev.4 (ESA, Paris, 1985).
10. HILLER, J. and TOLLISON R., Incentive versus Cost-Plus Contracts in Defense Procurement. *The Journal of Industrial Economics*, Vol. 26(3) (March 1978) p.239.
11. FISHER, I., An Evaluation of Incentive Contracting Experience. *Naval Research Logistics Quarterly*, Vol.16 (March 1969) p.63.
12. ANTHONY, R. and YOUNG, D., *Management Control in Nonprofit Organisations*. (Irwin, Homewood, 1984).
13. HOBAN, F. et al., *Readings in Program Control*, NASA SP-6103 (NASA, Washington, 1994).
14. KOELLE, H., The Transcost-model for Launch Vehicle Cost Estimation, *Acta Astronautica*, Vol11 (12) (1984). pp.803-817.
15. NASA, *Parametric Estimating handbook*, NASA, available under http://www.jsc.nasa.gov/bu2/hamaker.html (November 1999).
16. LARSON, W. and WERTZ, J., *Space Missions Analysis and Design*. (Kluwer, Dordrecht, 2nd Ed. 1993).
17. ESA, *Welcome to the Cost Analysis Division of ESA*, ESA, http://www.estec.esa.nl/eawww/ (March 2000)
18. GREVES, D. and SCHREIBER, B.: Engineering Costing Techniques in ESA, *ESA Bulletin* 81 (February 1995) pp. 63-83.
19. NASA, *Online Cost Models*, NASA, http://www.jsc.nasa.gov./bu2/models.htm (November 1999).
20. HUBER, R. and COHENDET, P., *Costing of Space Projects*, in HOUSTON, A. and RYCROFT, M., *Keys to Space* (McGraw-Hill, Boston, 1999) pp. 11-39 – 11-45.
21. HODGE, J., *Improving Cost Efficiency In Large Programs*, in NASA SP-6103 (NASA, Washington, 1994).
22. BARFIELD, J., RAIBORN, C. & DALTON, M., *Cost Accounting* (West Publ., 1991) pp.606-608.
23. INNES, J. and MITCHELL, F., *Activity-Based Costing*, (Chartered Institute of Accounting, London, 1990).
24. GOLDIN, D., *Statement before the United States Senate*, (23 April 1998).
25. PINKUS, R., SHUMAN, L., HUMMON, N., *Engineering Ethics: balancing cost, schedule and risk-taking. Lessons learnt from the Space Shuttle*. (Cambridge Univ. Press, Cambridge, 1997).

26. ADAMSON, J., *Privatization of Space Flight Operations*. in HASKELL, G. and RYCROFT, M., *New Space Markets* (Kluwer, Dordrecht, 1998), p. 140.
27. BROMLEY, D., *Reputation, Image and Impression Management*, (J. Wiley & Sons, Chichester, 1993). P.156.
28. AUGUSTINE, N., *Augustine's Laws*, (Viking, New York, 1986).
29. IN 'T VELD, J. and PEETERS, W., Keeping large projects under control: the importance of contract type selection. *International Journal of Project Management*, Vol.7 (3), (August 1989) pp. 155-162.
30. PEETERS, W.A., *The appropriate use of contract types in development contracts*. Ph.D. thesis at University of Technology Delft (the Netherlands), (1987). Also published as ESA STR-222, (ESA publ. Noordwijk, 1987).
31. IN 'T VELD, J., *Analyse van Organisatieproblemen*. (EPN, Houten, 7th ed.1998).
32. SLACHMUYLDERS, E., *Introduction to Space Mission Design*, in HOUSTON, A. and RYCROFT, M., *Keys to Space*, (McGraw-Hill, Boston, 1999).
33. ECSS, *Home Page*, under http://www.estec.esa.nl/ECSS/
34. ECSS, *Cost and Schedule Management Standard*, ECSS-M-60 (ESA, Noordwijk, 1996)
35. ECSS, *Risk Management*, ECSS-M-00-03 (ESA, Noordwijk, under preparation).
36. PREYSSL, G., ATKINS, R. and DEAK, K., Risk Management at ESA, *ESA Bulletin*, 97 (March 1999).
37. SOUDER, W. and BETHAY, D., The Risk Pyramid for New Product Development: An Application to Complex Aerospace Hardware, *Journal of Product Innovation Management*, 10 (1993) pp. 181-194.
38. DAOUPHARS, P., L'assurance des Risques Spatiales, in KAHN, P. *L'exploitation Commerciale de l'Espace*. (LITEC, Paris, 1992) p. 253.
39. THIEBAUT, W., *Space Insurance*, (ISU lecture, 1999).
40. RYCROFT, M., *The Cambridge Encyclopedia of Space*, (Cambridge Univ. Press, Cambridge, 1990), p.342.
41. DE SELDING, P., Reliability of New Launchers Stagnates, *Space News*, (May 29, 2000) p.1.
42. THIEBAUT, W., A New Approach to Insurance in the European Space Agency. *IAA-paper*, IAA-97-IAA6.2.07 (Turin, 1997).
43. GREEN, D., The Price Elasticity of Mass Preferences, *American Political Science Review*, Vol. 86(1) (March 1992) pp.128-148.

Chapter 5

DISTRIBUTION OF SPACE PRODUCTS

5.1 INTRODUCTION

Contrary to the concept in the business-to-consumer (B2C) market and the business-to-business (B2B) markets, few direct space products are physically distributed in the space environment that we are focusing on. The distribution of commercial space products will follow the traditional methods of Channel Distribution and will not be elaborated in view of the availability of many textbooks on this subject. Reference can for example be made to [1], presently a "standard work" in this field. As most of the elements evolving from the space sector are associated to the service marketing, this generic aspect will be treated briefly in the first section.

However, three aspects will be described and highlighted further, because they are closely related to the sector of space activities in general, namely:

1. Data distribution, or how are the data from space experiments brought to the end-user?
2. The Technology Life Cycle (TLC) and the innovation process.
3. Spin-off distribution, i.e. how the "cutting edge" products from space TLC developments are distributed to a broader community and how this process can be accelerated.

For all these aspects a number of elements are common and will be grouped in separate chapters, under

- The increased use of Internet techniques, and
- Intellectual property rights and patents

5.2 DISTRIBUTION OF SERVICES

Irrespective of the nature of the product, both tangible as well as non-tangible ones, the success of marketing depends on the requirement of easy accessibility. This is sometimes expressed as the "seven Rs" of logistics, namely [2] p.6:

> ***Ensuring the availability of the right product, in the right quantity and the right condition, at the right place, at the right time, for the right customer, at the right cost.***

In order to do so, the organisation will use a set of appropriate channels, a channel being defined as [3] p.460:

> ***A channel is a conduit for bringing together a marketer and a target customer at some place and time for the purpose of facilitating a transaction.***

This broad definition allows the identification of the different channels that are available for each different product independent of the nature of the product.

Examples can be found in literature as follows:

- A University can use different channels:

  - Centralised campus
  - Branches in suburban or other cities
  - An "Open University" structure with distance learning
  - Weekend concentrated courses (for post-graduate students having a profession)
  - Co-located courses in intensive "blocks", combined with self-study.

A decision will depend on the "target public"; research is more geared towards on-campus models, post-graduate business oriented education is more targeted towards remote education or "grouped" models.

- Analogous to Space Marketing is the example worked out in [4] p.740, on how the National Safety Council can reach drivers in order to for example point out the dangers of speeding in traffic. A distinction of channels is made between:

- Commercial entities, such as driving schools
- Service organisations, such as automobile clubs
- Schools
- Police
- Media (campaigns).

As we will see further on, a similar number of choices are available in Technology Transfer of for example space spin-off products. The information can be presented to the public by using a number of channels such as:

- Participation in fairs and trade shows
- Publications
- Internet sites
- Presentations to target publics
- Specialised Technology Transfer centres
- Use of commercial technology distributors.

Independent of the specifics of each product, a number of strategic problems apply to all channel decisions and they can be summarised, in analogy with [3] p.462 as:

- **Which quality of customer service do you offer?**

Some applicable examples can be given to illustrate this point: Brochures can be put on a table for distribution or handed out at a fair by an expert, ready to answer questions. A demonstration can be sent using videotape or shown physically. An effort can be made to invite a target group directly ad personam or a general press release can be made. It is evident that each service level will have a related price tag attached to it. Therefore, the organisation must evaluate how much value is put on "mass-channels" or more direct contact.

- **Do we use direct or indirect channels?**

Basically this refers to the use of one's own resources that has a number of distinct advantages (i.e. quicker response, better control over the messages, quicker awareness of potential problems). Lack of internal resources may lead to a decision to use intermediate channels. Often these intermediaries have the additional advantage of having a better knowledge of the market and more marketing experience.

- **Which intermediate channel organisations to be used?**

The organisation has to evaluate the proposals not only on the basis of the intrinsic advantages (such as knowledge of the market) but also as a function of length (one or more steps) and breadth (one or more in parallel) of the various approaches proposed.

- **How to achieve coordination and control?**

The basic problem is often that the supporting distribution organisation may have different perceptions and even different goals (such as distributing a maximum of leaflets with a minimum of effort). Controlling this may, in the end, absorb considerable resources. Typical methods to facilitate this are the use of requirements (for example in the contractual relationship) or rewards (if results can be correlated with a particular channel).

5.3 SPACE DATA DISTRIBUTION

Traditionally, space operations were centrally controlled via mission control and payload control centres. Scientists "brought" their products to the organisation and data were sent from space to these control centers, where they were picked up by the scientist (usually physically, such as copies on tapes and so on).

Since then, we have seen a very rapid change in communication methods. Using better links, such as ISDN, and making use of Internet techniques, even public telecommunication suppliers have made it possible for institutions with limited resources to have access to data in their laboratory. Recent ISS precursor missions, such as the 179 days EUROMIR95 mission, have taught the importance of this new evolution. For longer durations, the scientist cannot leave his office or laboratory. The only solution, which has been successfully experimented with, is to bring the data to the experimenter, NOT the experimenter to the data...

This has a number of distinct advantages:

- The experimenter can quickly analyse data and give feedback

- There are often laboratory models readily available to simulate off-nominal situations or remedial actions
- In the institution or laboratory all experts are on hand. Due to budgetary restrictions this cannot be expected in the control centre.

The terms telescience or remote science operations are commonly referred to as being the concept of remotely operating a scientific payload onboard of a space platform directly from the home location of the involved experimenter. The fundamental requirement to provide remote operations capabilities for an experiment platform or laboratory in space do, however, impose high safety, reliability, and security requirements on the implementation of the supporting communications system. These requirements were solved by the concept of the Interconnection Ground Subnetwork (IGS), which is an ESA provided autonomous Intranet with gateways to the secure networks of other space agencies. The main target is to provide an environment for scientists by rendering possible their virtual presence at the experiment workplace in space while remaining at their home site.

In recent years ESA has supported scientists to prepare for future remote science operations in space within the scope of precursor missions with Spacelab, Spacehab and the Russian space station MIR. An important feature of the "lessons learnt" from these projects was the awareness that communications carrier services are very expensive, when weighed against the cost for the experiment's development and operation. The future operational scenarios and the applied technologies have been described in [5], while [6] explains the activities at ESA to prepare for future high Quality of Service (QoS) communication implementations. These services within a secure operational Intranet are rather expensive, are not directly accessible by users with more moderate availability and reliability requirements concerning the communications services. To serve also these users the global Internet provides an attractive low cost alternative. Global public Internet technology and services owe their widespread acceptance partly to the basically distance-insensitive flat-fee tariffing concept. Thus the Internet appears to be the most straightforward and low-cost solution in order to deliver experiment data and video streams even simultaneously to a very large user community.

In the ISS era (figure 5.1), telemetry data, i.e. scientific and house keeping data and the video streams will be transferred to the ground using a network of geo-stationary Tracking and Data Relay Satellites (TDRS). The antennas of the NASA ground station in Whitesands will terminate the space to ground link. The data then will be propagated via high-speed communication lines to the Marshall Space

Flight Centre in Huntsville, Alabama, where the Payload Operations and Integration Centre (POIC) and the Payload Data Service System (PDSS) are located. Here the European science telemetry data will be transferred to the ESA relay and transported to the IGS central node at ESA in Darmstadt, Germany, using a high reliable Asynchronous Transfer Mode (ATM) service over a trans-Atlantic link. The IGS central node containing the major operational communications and network management infrastructure is the centre of a star-like network, interconnecting all involved European operations entities.

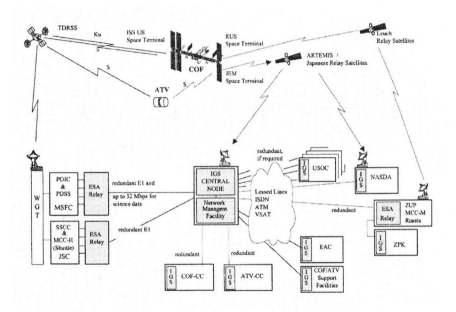

(Courtesy: U. Christ/ESA)

Fig. 5.1: Overview of ISS dataflow

The requirements for interactive operations with experiments in space are the following:
- Most users need to monitor their experiment by receiving the telemetry data. They need to display the received house keeping and experimental data on a monitor, preferably at their home site
- The second requirement, the on-line availability of data, is decisive for interactive remote operations. Here distinction is made between on-line and off-line data distribution. On-line implies to receive the data nearly in real time (i.e. within seconds up to a few minutes). Off-line stands for data distribution of collected science data with a significant time delay, e.g. over night

- This leads to the third requirement, the experiment control. If the facility design provides this capability, the user should be able to control his experiment himself. As he is most familiar with his experiment, he should be able to directly issue telecommands
- In view of the longer time lapses between experiments, the user wants to stay at home. In the past the user had to move with most of his ground support equipment to a NASA site to monitor his experiment. Now the user can operate his experiment from his Institute, where he has his supporting science staff and all the necessary processing resources available.

The users do not only wish to receive a feedback but also want to have an influence on when their experiment will be executed. This can be highly important for industrial users who do not always have absolute freedom for their researchers to be available, but who also may use the PR-effect of such experiments in order to invite an audience. The key to this problem is the operations planning which will have to take into account the sharing of on-board resources by the partners, within the allocated resource budgets. The planning foresees coordination at different levels (strategic, tactical and increment planning) [7].

Whereas the partners will communicate technically via dedicated servers, further information distribution will be done using common modern communications tools such as the WWW. The principle of data distribution chosen is "information casting", whereby all requesters will receive information about their resource request, including potential resource conflicts and indications for conflict resolution. One aspect of the real-time data concept should not be overlooked. Whereas the delays in communication are within certain acceptance limits for Low Earth Orbits, they increase with increasing distances to a level where real-time telescience for rapid evolving processes become virtually impossible and will require direct human presence, as can be seen from table 5.1 [8], p.876

DESTINATION	DISTANCE (KM)	ROUND TRIP DELAY (S)
Low Earth orbit	500	3×10^{-3}
Geostationary	35,700	0.24
Moon	384,000	1.28
Mars (average)	227,000,000	757

Table 5.1: Round trip propagation delay as function of destination

5.4 INCREASED INFORMATION SPEEDS: THE TLC CONCEPT

In the last few decades, authors have pointed out that the traditional Product Life Cycles (PLCs) have changed considerably, particularly in technology-related areas. This change is described in [9] as:

At the heart of the current change is the fact that Product Life Cycles are getting shorter. Every industry we look at seems to be undergoing shorter cycles. One moment, infancy; the next moment, old age. Blink and you miss the market...

A lot of these changes are due to rapidly changing technology. Whereas in the past it was feasible to replace for example computers every 7 years, they are now old-fashioned within 3 years due to faster and more powerful components. This evidently means that new technologies have to reach the markets faster than in the past and need to be put in production faster, also called "Rapid Ramp-up" techniques. The introduction of new products did not always happen so quickly after its invention. In table 5.2 some examples are given from the past [10]. In the past and even nowadays obvious technological projects sometimes take a long time to come on the market; the well-known zipper was patented in 1891 and only produced in 1918, 27 years later...

PRODUCT	INVENTOR	YEAR	INNOVATOR	YEAR	DELAY (YRS)
Television	Zworykin	1919	Westinghouse	1941	22
Wireless Telegraph	Hertz	1889	Marconi	1897	8
Xerography	Carlson	1937	Haloid Corp.	1950	13
Radar	Marconi	1922	Radio Electricité	1935	13
Jet Engine	Whittle	1929	Rolls-Royce	1943	14
Records	Goldmark	1945	Columbia Rec.	1948	3
Plexiglas	Chalmers	1929	Imperial Chem.	1932	3
Freon	Midgley	1930	Du Pont	1931	1

Table 5.2: Examples of lead time to production in the past

A systematic overview is reported in [11], whereby the historical time between invention and first application can be plotted as per figure 5.2.

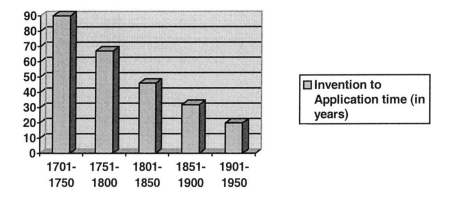

Fig. 5.2: Evolution of time between Invention and first application (in years)

We can note from the figure that the average time, required to bring a new invention to the market, has been nearly-linear decreasing over time (although evidently an asymptotic value will be reached in the future).

This brings us to the topic of Technology Life Cycle (TLC), which is gaining more importance over the previous concept of the PLCs in an increasingly number of areas (as also many consumer products use more Information Technology and therefore are more technology dependent). Popper and Buskirk [12] defined the TLC as the evolution of Technology through a marketplace and introduced the relation between TLC and marketing, by dividing the Technology Life Cycle into six basic phases:

1. **Cutting Edge**
This refers to the level of technology, which is ahead of even the most sophisticated applications in the marketplace, thus more closely related to research than to development. Cutting edge organisations in general are less interested in marketing their developments.

2. **State of the Art**
State of the Art (also called SOTA) companies specialise in adapting developed cutting-edge technologies to market needs and applications. In some cases they are Cutting Edge organisations which have developed

further. A typical example are the "Technology Valleys" that are often found around leading universities and which are still working together closely with those universities.

3. Advanced Stage
In this phase the operating mentality shifts to successfully marketing the product. In most cases the technology is "sold" by the previous types of organisations (e.g. as a patent) to a more marketing oriented company.

4. Mainstream
At this stage the market is fully developed. The product is standardised and the strategic focus is placed on low-cost production. Companies in this phase will need considerable production resources; production is often shifted to countries with lower labour costs.

5. Maturity
Competition shifts to customer service as prices stabilise. The product is now more a technological commodity and traditional marketing techniques are used, trying to find new distribution channels and product modifications, combined with (price-based) sales promotion.

6. Decline
Other technologies replace the declined technology. The firm can only continue to use the original technology for a while by pricing itself substantially below the newer technologies.

This phasing is very important for organisations, as it clearly shows the difference in orientation in organisational culture for each phase. Organisations have to decide in which part of the TLC they want to operate. Many companies have tried to "grow" with the technology phase but then were confronted at a certain stage with a market that they were not able to manage. Some important differences per phase are presented in table 5.3

If we apply this to our space activities

- The basic technology programs initiated by the agencies such as ESA, CNES, and DLR do a lot of "cutting edge" work
- A number of space companies are involved in this by the execution of contracts under the technology framework and "sell" the technology that they acquire to mainstream companies.

Variable	Cutting Edge	State of the Art	Advance	Main-stream	Mature	Decline
Product	Feasibility	Specialised	Segment	Standard	Shared	Obsolete
Price	Very High	High	High, price war	Market price	Market price	Dumping price
Distrib. channels	Direct sales	Sales network	Multiple channels	Market channels	Reduced channels	Limited
Promotion	Personal selling	Trade shows	Distrib. Network	Brand Name	Sales Promo	Limited
Technology	> 95%	→			< 5%	

Table 5.3: Interrelation between Marketing and the TLC

A special emphasis is put by ESA in stimulating Small and Medium Enterprises to implement the technologies they acquired by their participation in space activities, in particular through technology programmes. This effort is framed in the TLC philosophy whereby in the "State-of-the-Art" stage of a new technology, there is still a very limited interest from the mainstream companies but, on the other hand, there may be an interesting market for the small, dynamic "Silicon-Valley" type of companies. [13].

Examples to illustrate the above can be found in [14], such as:

- Fuel cells were examined in ESA for use in the Hermes space transport vehicle (cutting edge phase). Dornier has been working under ESA contract on safe, reliable fuel cells for use in spacecraft (State of the Art Development). In 1996 Dornier became part of Daimler-Chrysler, and Mercedes saw in particular an interest in marketing the Necar (New Electric Car) on the basis of the current A-class automobile (Advanced and initial Mainstream Development)
- ESA envisaged developing a new spacesuit. As part of this, a cooling mechanism needed to be developed, based on Peletier elements and light rechargeable battery packs. Initial work was done with a Belgian University and a small Belgian and a Spanish company (Cutting edge). A Spanish company, Zodiac, developed the prototypes (State of the Art development) and realised that this technique could be used in a broader application. Zodiac has developed the use for bullet-proof vests and for firemen (Initial advanced development)

- CNES developed a software package for radar satellites, such as ESA's ERS-1, to monitor environmental changes on Earth (Cutting edge). Realising a broader use, CNES then adapted the Software package, called Diapason, for geophysical laboratories (State of the Art phase). Presently, industrial companies are interested in licenses of the package for commercial use (Advanced phase).

In order to illustrate that these trends are effective, a McKinsey team [15] interviewed a number of technology companies worldwide and came to the conclusion that leading companies:

- Commercialise two to three times the number of new products and processes as do their competitors of comparable size
- Incorporate two to three times as many technologies in their products, and
- Bring their products to market in less than half the time.

It is important to note that Marketing and Technology are not a priori considered as "walking hand in hand". In fact we can note a number of conflicting interests in the course of the product innovation process, as we can see from the examples in table 5.4

MARKETING INTEREST	TECHNOLOGY INTEREST
Concentrate on Product Development	Concentrate on Research
Develop and control specifications	Try out different technology paths at the same time
Freeze design as quickly as possible	Keep options open
Accelerate Time-to-Market	Do extensive pre-testing
Rapid Ramp-up (to go in production)	First start a "pilot-plant"
Enhance Customer Acceptance	Emphasise technological features
Keep proprietary rights	Publish results

Table 5.4: Examples of conflicting Marketing/Technology Interests

A well-known example on "Rapid Ramp-up" in this context can be found in the software market. Driven by strong competition marketing oriented firms put new products on the market even when they know that there are still "bugs" in the software. In fact they use the market virtually to debug their software and use the feedback to improve their products in the new versions. The technique has been heavily criticised, even in-house by software firms, but has proven to be a win-win strategy.

A lot of research is carried out on the conflicting cultures between R&D and Marketing staff. Based upon the results of various studies Griffin [16] even warns about the danger that the differences, as presented in table 5.5, may become so strong that both groups become self-contained "societies" in the organisations in which they reside.

FACTOR	MARKETING	R&D
Time Orientation	Short	Long
Professional Orientation	Market	Science
Project Preference	Incremental	Advanced
Organisational Orientation	Medium	Low
Bureaucratic Orientation	More	Less

Table 5.5: Cultural differences between Marketing and R&D Staff

Capon and Glazer [17] tried in 1987 to better merge both interests. Their main message was that integrating a technology and a marketing strategy are paramount elements that effect corporate success in the present, rapid changing environment. As a technique they propose that firms "inventorise" their technology portfolio, in comparison with competitors (benchmarking) and derive from there onwards the market segment or phase of the TLC they want to develop. This is in fact the answer to the old "Which business are you in?" management question. Moenaert et al. [18] have, in the same context, context, studied the communication between R&D and Marketing. One of the metaphoric statements in the article

I have a DREAM: Development by Research, Engineering And Marketing...

Is very clearly emphasising at the same time the problem as well as the solution. The main finding of the study is that communication between marketing and R&D is improved by:

- Formalisation of the projects (for example a clear "project plan")
- Decentralisation (i.e. open availability of the project information)
- Positive interfunctional climate (mutual interest and trust in each other's activities)
- Role flexibility (different tasks in the course of a project, such as an R&D expert talking to the customers).

Based upon statistical analysis, however, only project formalisation and the quality of the interfunctional climate were found to have a significant effect on project success

5.5 THE INNOVATION FLOW

Very similar to the TLC approach is the "Innovation Model" approach. [19]. A traditional innovation model can be represented as per table 5.6.

INPUT	Public Funds	Public Funds, minor industrial funds	Industrial Funds, minor public support	Industrial Funds
PROCESS	Basic Research	Applied Research	Development	Production
OUTPUT	Publications	Patents	Production skills	New Products

Table 5.6 Schematic Innovation Process

The correlation between tables 5.3 and 5.6 is considerable, only in the first case is technology the driving parameter, whereas in the latter case the innovation process is used as main element. We can, however, associate:
- Cutting Edge with Fundamental Research
- State-of-the-Art with Applied Research
- Advanced phase with Development.

In the concept of Innovation, more emphasis is put on the different categories of R&D. Indeed, joining the concept of Research and Development in this one acronym induces a potential problem that both get mixed up in high-level presentations. As one example, there are data found easily on the financial value of R&D in total, but rarely on the R and the D values separately.

One of the figures on the difference is reported in [20] p.140 and illustrates the considerable distinction: whereas development spending accounts for 62% of all U.S. R&D activities (1998 figures), 75% thereof is done by industry. On the other hand basic research counts for 15% of U.S. R&D spending (the remainder is spent on intermediate areas such as applied research), and only 4% is associated to industrial activities. For the sake of clarity, the OECD definition are recalled as:

__Applied Research__ is the process of discovering new scientific knowledge, which has the potential to act as a platform for the subsequent development of commercially viable products and manufacturing processes.

__Development__ is systematic work, drawing on existing knowledge gained from research and practical experience, that is directed to producing new materials, products and devices; to installing new processes, systems and services, or to improving substantially those already produced or installed.

This in contrast to **Basic Research** which can be defined as:

Experimental or theoretical work undertaken primarily to acquire new knowledge of the underlying foundations of phenomena and observable facts, without any particular application in view.

The differences between Basic Research, Applied Research and Development can be best presented as per table 5.7, [21] p.142

	BASIC RESEARCH	APPLIED RESEARCH	DEVELOPMENT
Degree of orientation	Minimal	Medium/high	Medium/high
Commercial objectives	Low	High	Medium/high
Payback Criterion	Long-term	Medium-term	Short-term
Operational time horizon	Long-run	Medium-run	Short-run
Degree of uncertainty	High	Medium	Low

Table 5.7: Characteristics of R&D elements

The absence of commercial objectives and the long-term orientation underline the "public aspect" of basic research and explains the low industrial participation (moreover, it would be difficult to support such level of research in a competitive environment). On the other hand only basic research intends to contribute to the advance of fundamental science and is therefore an essential part of societal long-term development. Let us now consider the consequences of these concepts in the framework of the Innovation Process.

In the 1960's, the **"Technology Push-model"** was adhered to in literature, whereby it was assumed that the results from fundamental research would "push" and motivate new products in analogy with similar models in production processes [22]. In other words, the belief was that increased economic development could be achieved by increased funding for research. Later, more emphasis was put on the **"Market Pull"** effect, whereby the market demand for new products stimulates the innovative cycle and would generate research funds.

It should not be a surprise that both models were linked together in the 1980's by an integrated model, also called the **"Chain-link model of Innovation"** [23]. This interactive model constitutes a linking of research and the market with a number of feedback loops.

This approach attempts to satisfy the needs of technology transfer by approaching the problem from both ends, i.e. using push and pull at the same time. It originated mainly in Japan where it is successfully implemented inter alia in the automobile industry (for example Toyota). An important feature of this model, for our specific case, is the existence of an accessible Knowledge Base for all the parties concerned and a permanent interaction with that Knowledge Base.

For completeness, we have to mention here that a new generation of innovation models is presently emerging, namely the **"System Integration and Networking (SIN)"** models. The basic feature is the central use of a sophisticated computer toolkit, in order to enhance the speed and efficiency of product development across the whole system of innovation. Therefore it can also be described as a "lean information process". Typical of this approach are the use of data sharing systems, expert systems and CAD/CAE techniques (to reduce testing time and costs). Very specific is the network approach, linking CAD systems of prime suppliers and datalinks to (outside) R&D collaborators. [24] pp. 42-50.

The McKinsey consultancy firm came to the conclusion that reducing speed-to-market has a greater impact on product profitability than making similar reductions in either investment cost or R&D expenses. Therefore mainstream companies often follow as ground rules [25]:

- In order to acquire substantially new know-how, make a deal on equity-basis (e.g. joint venture)
- In case of complementary innovation, procure on non-equity-basis (Buyout, license)
- For incremental innovation, prefer development in the in-house R&D department (also to avoid know-how transfer).

The increased TLC speed drives companies to such a "spin-in" approach, as they realise that time is too precious to try to rely on own innovations only. This effect is more applicable to earlier phases of the TLC; a typical example has been the general request for spin-in made by NASA before developing ISS technologies. Spin-in can be schematically presented as per figure 5.3

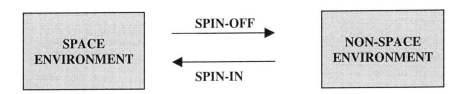

Fig. 5.3 Spin-off versus Spin-in

A side-effect of this is a higher degree of specialisation, whereby the "State-of-the-Art" firms are not seen as competitive but more as complementary by the larger mainstream space companies.

We should not forget that the space sector is more oriented towards using and combining technologies and less towards generating this. Other sectors often spend more resources on R&D and therefore have broader technology background. A part thereof can be used in the space activities, provided the right information channels are maintained.

If we compare 1996 figures, as published in [26] p.177 and rank the R&D expenditure across industry sectors, we note from figure 5.4 that aerospace activities score lower than we would expect, even lower than the "General Manufacturing" sector.

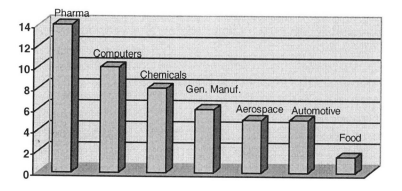

Fig. 5.4: R&D expenditure as % of sales across industry sectors

Also on a global scale the differences are considerable according to OECD statistics [26] p. 178; for the early 1990's the comparable figures of R&D expenditure were:

- 1.96 % of the GNP for the 12 EU countries
- 2.74 % of the GNP for the U.S.
- 2.87 % of the GNP for Japan.

These figures clearly illustrate the danger of a growing "technology gap" between Europe and other major industrial countries.

5.6 SPIN-OFF DISTRIBUTION

In the specific context we are dealing with we refer mainly to the distribution of data. The rapid increase of (electronic) communication means have opened completely new frontiers in this respect.

Originally NASA worked with a Technology Transfer System providing geographic coverage all over the U.S., primarily in areas of industrial concentration. This was divided into a number of Field Centre Technology

Utilisation Offices, which were in fact regional offices to give support as follows:

- Industrial applications centers, more focused towards information retrieval and assistance to users
- Computer Software Management and Information Centers (COSMICs), where computer programs could be made adaptable to secondary use
- Application teams, working with institutions in order to use NASA technology to solve public problems.

The aim was clearly to bring the technology closer to the potential users, hence the "Technology Transfer" aspect. These evolved into the 11 Centres for the Commercial Development of Space (CCDS) co-located with a number of universities and research institutes across the U.S. Detailed information is accessible via:

http://nctn.hq.nasa.gov/innovation/Innovation23/CCDS.html

With the arrival of Goldin, this task was boosted in the "Office of advanced concepts". The main objectives being described [27] as:

- Bridging the gap between NASA technology development and real-world applications
- Provide a point of contact for those who want NASA's help
- Accelerate the pace of technology transfer to the commercial sector
- Stimulate commercial activities.

A major emphasis is put on creating an "Electronic Commercial Network", which can be accessed inter alia via

http://nctn.hq.nasa.gov/

Whereas a complete "toolbox" for industry is available under

http://technology.jsc.nasa.gov/

A similar approach was taken by ESA, specifically in the Microgravity area, by creating a number of user support centers, called MUSC (Microgravity User Support Centers). The underlying philosophy was developed as early as 1992 and is now implemented in the framework of the ISS activities in Europe.

Another initiative in this area is the RADIUS project initiated by ESA [28]. The basic approach here is to identify leading scientific groups working in well-defined areas of research using Microgravity and to provide them with the means for promoting industrial research in those areas by lending high-quality scientific support to industrial partners. Putting the instruments directly in the hands of scientists is intended to improve communication. The implementation is promoted via Spacelink Europe; an economic interest group specialised in technology transfer.

The latest version of the ESA Technology Transfer Programme, called IMPACT [29], classifies the spin-off products In a number of categories as presented and illustrated in table 5.8.

Based upon the fact that space products are designed to cope with demanding conditions, an inventive approach is ESA's "Harsh-Environment Initiative" [30], specifically operating in Canada and Northern Europe. Specific target groups are areas invaded by sea ice, offshore oil & gas exploitation and mining in remote areas. The rationale behind these target groups is the fact that space products, designed to withstand very severe conditions (for example high temperature gradients) could be applied in such hostile environments. Some typical examples of this spin-off are:

- Use of "ice-phobic" materials to protect microwave communication towers in Labrador
- Monitoring of ground motion using satellites, for the monitoring of unstable slopes in which pipelines have been installed (in order to prevent pipeline failures)
- Tele-operation of drilling equipment in hazardous environments, using space-robotics software and control tools.

Under the subject "proactive approaches", we have to mention the initiative to bring together the European plastic industry and ESA. Again the rationale behind this initiative was that strength, adaptability and durability of lightweight plastic films and polymer-based fibres have made them a key component in space vehicles, equipment and clothing. The joint report [31] is an excellent example of opening an active "channel" between both communities.

CHAPTER 5 : DISTRIBUTION OF SPACE PRODUCTS 171

CATEGORY	EXAMPLES	APPLICATIONS
Materials	• Electromagnetic Shielding Coatings • New Anti-static Paints • Composite products (CFRP, CMC)	• Microcomputers / Medical • Safety / explosions • Furnaces, mechanical parts
Computer hardware/ Software	• Multimedia Archiving • Industrial Software Engineering • Simulation Language	• Audio-visual field • Safety critical software • Process industry
Automation/ Robotics	• High-accuracy pointing systems • Visual monitoring camera	• Robotics/ manufacturing • Hazardous environments
Sensors/ Measuring	• Porous-Silicon Colour sensors • Pinch-force dynamometer • Trace Gas Analyser	• Computer peripherals • Physiotherapy monitoring • Safety (workplaces, planes…)
Mechanical Components	• Piezoelectric Actuators • Capillary Cooling Loop	• Optics / micro-electronics • Computer cooling
Precision mechanics/ Optics	• Electroformed optical mirrors • Cryocooled seals	• X-ray equipment • Heart pumps
Communication	• RF remote control • Fibre-optic video transmission	• Process industry • Audio-visual applications
Electronics/ Opto-electronics	• Data compression • Rigid-flex circuit boards	• Radar signals • Load-bearing constructions
Life Science/ Medical	• Environmental protective clothing • Organic waste recycling	• Petrochemical industry • Environmental technology

Table 5.8: Categories of ESA spin-off orientations

An ESA Internet site on Technology Transfer has been established under

http://WWW.esa.int/technology

and, in analogy with NASA, one can note a shift from classical publications to Internet documents in this rapid changing environment.

(Photo: ESA/ P. Brisson)

Fig. 5.5: Examples of Spin-off Distribution

5.7 THE USE OF THE INTERNET

THE INTERNET

As we move into the new era of The Information Age, the expectations of consumers are being radically altered. The Internet, more than any other technology, is teaching people that they can go online and quickly find out about any subject of personal interest. Not only are they able to obtain a substantial depth of information, they are able to do this instantaneously. Consumers in the Information Age are no longer satisfied with requesting information and then awaiting its arrival. They expect instant information satisfaction and they are learning quickly that the Internet is the way to get it.

An important aspect of the Internet is the vast number of people who are now connected. Also the commercial value of the Internet is increasing according to predictions of the growth of Internet generated revenue.

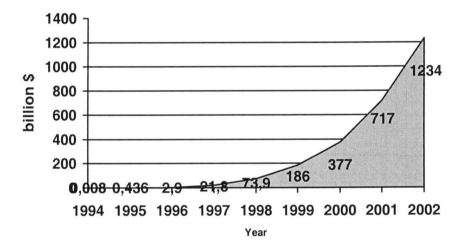

Source : EUROTROTTER.com

Fig. 5.6 Worldwide Internet Generated Revenue growth (in billion $)

Although figure 5.6 shows the worldwide Internet Generated Revenue, it also teaches us about the expectations of the Internet in general. A survey of Yahoo! Life (one of the leading Internet search engines) pointed out that the total number of users on the Internet was about 256 million in October 1999 and is still growing day by day. A forecast of Nua Internet Surveys speaks of

350 million users by the end of the year 2005. Of these 256 million there are 95,5 million users within the ten leading European countries.

It took radio 30 years to reach 90 million people. Television took 15 years to achieve a similar market penetration. The Internet has achieved this in just three years and its rate of growth continues to climb.

Therefore the possibilities in this area are just beginning. One of the potentially successful mechanisms will be "database marketing" probably. Under this technique an organisation will be able to [32]:

- Select an individual target group via databases (whereby several criteria, such as age, educational level, race and so on can be used)
- Develop specific programmes for that group
- Cost-effectively direct electronic mail to them
- Track and evaluate the results.

This could also be used to form a "closed loop of information" with certain customers; i.e. responses can be collected and analysed as to consumer's preferences, enabling an organisation to try out "cause-and-effect" relationships.

These techniques require enormous amounts of dataprocessing, up to the individual customer level. However, this type of data handling is becoming increasingly cheaper via the Internet techniques, which are increasingly able to determine user characteristics and behaviour. One can easily forecast that this trend will continue and will offer unprecedented solutions for the physical distribution problem. It can be forecast that projects such as the "Teledisc" of Bill Gates will accelerate this process exponentially.

Many ESA data and information are already available on the WWW or Internet but NASA is certainly more advanced in this area. Knowing that the Mars Pathfinder results would interest the general public they installed 25 WEB servers to cope with 35 million daily requests. It goes without saying that this boosted NASA's image considerably [33]. NASA's preference for the Internet as a channel is based certainly on the in-house evaluation as per table 5.9 [34]:

CHAPTER 5 : DISTRIBUTION OF SPACE PRODUCTS

	PERSONAL	PRINTED	BROADCAST	INTERNET
Portability	0	High	0	High
Tunability	High	Low	High	Low
Cost/ Impression	High	Low	High	Low
Ease of creation	Easy	Medium	Difficult	Medium
Delivery time	Minutes	Hours	Instantaneous	Instantaneous
Density	Verbal	300 words/p.	Multimedia	Multimedia
Flux	2-3 words/sec	5 words/sec	Multimedia	Multimedia
Lifetime	Seconds	Days/years	Seconds	Days/years

Table 5.9: Channel comparison of information

From the table we can conclude:

- Personal contacts are effective and can be easily tuned (direct feedback) but are very costly (hence only to be used for a limited target group)
- Printed information is cheap and "stays", but is difficult to adapt to a specific group
- Broadcasting is difficult to implement and expensive, but offers very good communicative possibilities
- Internet is a good compromise, being relatively inexpensive and printable if needed, with full multimedia possibilities.

Based upon this perception, in 7 December 1999 NASA published an announcement "Notice of Intent to Negotiate Partnership Agreements for the Development of Multi-Media Products and Services Related to the Exploration and Development of Space". From the 12 offers received, the one from Dreamtime Holdings, Inc., a Silicon-Valley company specialised in hi-tech multi-media was selected. The plan is to develop a space Internet Vertical Portal (a "Vortal"), a multimedia database and television and documentary programming. NASA will put its archives at the disposal of this media and will provide high-definition television coverage of astronauts onboard the International Space Station.

The first step will be the creation of a website featuring the latest in interactive technology, by combining video, audio, still photographs, high-resolution images, historical documents and technical information.
Progress can be followed under

http://www.dreamtime.com

Equally important is the perception of people on the quality of the various channels; empirical research teaches us that [35]:

- The Web site is excellent for conveying information and detail (scoring similar to direct mailing brochures)
- The Web site is seen as cost-effective (important in view of the potential criticism against nonprofit marketing)
- The Web site is considered as a rational medium
- The Web site is effective in precipitating actions (less than direct contact, but better than e.g. television)
- The Web site is effective for both short- and long-term objectives.

On the other hand:

- The Web site is not effective for stimulating emotions (contrary to television, the Web site is seen as a "cognitive medium")
- The Web site is a less-effective medium for attention-getting devices
- The Web site is less effective for changing and maintaining attitudes.

Due to the emphasis on the cognitive aspects, Web sites are an excellent tool for space technology transfer and marketing channels (see for example figure 5.7). For promotional and communication aspects, it is a considerable tool but needs support from other distribution channels specifically when emotions are targeted.

One unusual application of Internet in space activities is the search for flight opportunities [36]. NASA created the Access To Space (ATS) group in 1998 to help scientists find better and more affordable access to space. It is based upon NASA's mission database and accessible via the WEB for the NASA scientific community trying to allocate a flight opportunity for their experiments (such as the availability of "Get Away Specials" on board of the Shuttle). It is reported that, after recent amelioration, some 1000 visitors monthly are recorded on the site. This, and similar applications, demonstrate the vast possibilities of using the Internet as a channel.

Fig. 5.7 WEB site of the German space agency DLR (www.dlr.de)

5.8 INTELLECTUAL PROPERTY ASPECTS AND DISTRIBUTION

DATA PROTECTION

New products have two sides; on the one hand they open new markets, on the other hand they attract competitors trying to imitate the product. Patents are a classical method of protecting against competition, but they are of limited protection only (and can easily be circumvented). An additional problem for public funded organisations is a limited right to protect such Intellectual Property, because it is the property of all the taxpayers.

Intellectual Property rules are most advanced in the fields of data collection and distribution in the Remote Sensing and telecommunication fields, because the commercial publications have "pushed" legislators to come up with solutions for this. However, presently the problem is not a lack of regulations, but rather a lack of uniform regulations worldwide. Specifically in the area of data protection different approaches are noted by Clerc [37]. The concept of data policy can be subdivided in two aspects:

- Data Reservation (data ownership and legal data protection)
- Data Distribution (includes pricing policy and competition regulations).

The most common way of data protection is Copyright Protection, which can be defined as:

> ***Copyright is an intangible property right, which grants an author the exclusive right to his original creation. The author can transfer this right to a third party, e.g. by issuing a reproduction license.***

Copyright protection is relatively simple and is transferable internationally ("Berne Convention"). On the other hand, the contents of the raw data are not protected. Therefore, other organisations prefer

- The "Data Base Protection" solution (as it is based upon a EU directive)
- Technical Protection (such as encryption)

ESA for example has a preference to follow the "Data Base Protection" approach [38], mainly for the reason that in this way also the raw (satellite) data can be protected.

The overall problem lies in the fact that the only International legal instrument widely accepted at present is the so-called Outer Space Treaty, which provides the following principles [39]:

a. **States have the right to use outer space**, but not to appropriate it.
   Outer space is considered as common property ("res communis") whereby every State is free to undertake activities but may not obstruct the same free use by another State.
b. **States have taken the obligation to use outer space for peaceful purposes.** The Treaty further specifies that, on the one hand, celestial bodies shall only be used for peaceful purposes and hence cannot be militarised; on the other hand, there is a prohibition against placing mass destruction weapons in space orbits (Note that this does not exclude the deployment of non-aggressive military systems such as reconnaissance satellites).
c. **States are responsible for and have to supervise private activities**.
   In order to assure compliance with the Treaty, States therefore must give an authorisation and continuously supervise non-governmental

activities of companies registered in that State which intend to carry out space activities.

d. **All space objects must be registered**. Registration is carried out in the international register kept by the Secretary General of the UN, indicating designation, functions, orbital parameters, frequencies used and launch date of each space object.

e. **States retain jurisdiction and control over its space objects**. While an object is in outer space, the State of registry retains the jurisdiction over this object.

f. **Assignment of Liability for damage**. This aspect is further covered by the "Convention on International Liability for Damage Caused by Space Objects" [47]. In case of damages to a third party, the "Launching State" is absolutely liable for damage occurring to objects or persons on Earth or planes. The notion of the **"Launching State"** is very important in this context and defined as:

> *The State that launches, procures a launch or from whose territory or facility the space object in question is launched.*

g. **International Law is applicable to space activities**. The Treaty specifies that activities in Outer Space shall be carried in accordance with international law, including the Charter of the United Nations.

This brief summary teaches us a number of important consequences:

- The Treaty is based upon space activities from Governments primarily (see the Launching State concept) and is not fully compliant with for example the concept of private launching entities
- Commercial activities by private organisations are not properly covered, specifically due to the "free use" provision of outer space (excluding property rights)
- The Treaty focuses on single State endeavours and is not compliant with the trend of global and transnational endeavours such as the SeaLaunch project
- From a Liability point of view, national laws still seem to be needed under the terms of this Treaty.

We should not forget that the Treaty was conceived in the 1960's, in a completely different political context. Moreover, at the point in time when drafting the Treaty, not in the wildest imagination were commercial applications of space activities by non-governmental entities taken into consideration. An analysis of further potential steps is provided in [48] and foresees the following possibilities:

1. **New Space Regulation on National Level.**
Such national regulation is already existing in the U.S. and Russia, and, in Europe, in the U.K. and Sweden. Therefore, extending this to other countries would be relatively simple; however, it will not solve the transnational endeavours. On the other hand, such national regulations are needed anyway in view of the Liability aspects.

2. **New Space Regulation on European level**
Although this would be a next logical step, it would only lead to a European Commission (EC) Regulation or Directive harmonising the relations between the EC countries and hence, it will not contribute to the global problem.

3. **New Space Regulation on International Level.**
Basically this would represent the best solution but at the same time, it is the most difficult one to be achieved. It would, in fact, introduce the concept of private law in the space environment [49] and therefore, deviate from the International Treaty approach (but at the same time introduce more familiar legal principles, which are well-known to all parties). One of the potential candidates to play a role in such a process is the World Trade Organisation (WTO), established in 1995. Indeed, the WTO established a legal framework for commercial telecommunication services in 1997, thereby covering 93% of the commercial telecommunication activities, based upon the 69 States, which have committed themselves by signing this agreement [49]. Another action taken by the WTO, on the TRIPS Agreement (Trade Related Aspects of Intellectual Property Rights) [50] could equally contribute to the harmonisation of Intellectual Property aspects on a global scale. Alternatively, modifications to the existing conventions could also be considered as a potential instrument. This could be done by protocols or by Amendments to the signed Conventions.

4. **A Code of Conduct Approach.**
This is an alternative solution and rather a complementary one to the previously mentioned and more basic solutions. Under such Code of Conduct, the mutual recognition of intellectual property rights of countries and the companies involved could be provided. This approach, of course, does not have the same legal value but will be easier and faster to adopt. In

the case of the International Space Station it may turn out to be the only possible timely solution.

In this respect also we have to mention a potential conflict on legal and contractual provisions with the previously mentioned PPP concept. Full data dissemination would, of course, block the competitive advantage of the interested industrial party, and therefore, compromise solutions that need to be established [45]. The main differences between a "traditional" Agency project and a more flexible PPP approach is given in table 5.10.

ASPECT	TRADITIONAL AGENCY APPROACH	PPP APPROACH
Financing (LCC)	100% Agency financed	Max. 50% financed
Requirements	Agency	Industry, in view of commercial interest
Statement of Work	Agency	Agency (only top level)
Programme of Work	Agency	Industry (mainly)
Contract Conditions	Agency	Tailored

Table 5.10: Adaptation of contractual instruments for a PPP approach

In the contract conditions, the full protection of results for the financing contractor have to be guaranteed, with a limited number of Agency rights (such as free licensing rights).

PATENTS

As a formal definition,

A patent is a document, which grants its owner a legally enforceable right to exclude others from using the invention.

In our specific context one has to stress the point that an invention is only patentable if it meets following substantive conditions:
- It must be new
- It must involve an inventive step
- It must be industrially applicable.

Specifically the third component, the industrial applicability, is often difficult to substantiate in the case of space activities, and industrial spin-off activities, are in general easier to patent. Requesting a patent is a lengthy process and can be very expensive if a patent is requested for many countries. Most patents in the European context are delivered by the European Patent Office (EPO), which maintains a database of over 30 million patents. Specifics on patents and requests for patents can be found under

http://www.european-patent-office.org

Closely linked to the concept of a patent is the license, defined as:

A license is a means to commercially exploit a patent by giving someone else the authorisation to use an invention

In a first instance, the use of patents may look to somehow conflict with the objectives of space agencies. The rationale is explained in Kallenbach [46], who distinguishes between ESA in-house inventions and Contractor inventions. ESA patents are regularly published in the "Catalogue of ESA Patents" [47], as well as in a quarterly newsletter, called "Preparing for the Future". In the case of inventions made by ESA contractors in the course of executing work under contract, ESA has adopted a license policy, according to which the contractor keeps the rights in such inventions, but grants royalty-free irrevocable licenses to ESA and to its Member States.

This policy of free licenses proved to be a drawback in a competitive market situation. The influence of regulations on patents and licenses is considerable and one of the important factors of an environmental influence on hi-tech activities. This is illustrated by the effect in the U.S. of the Bayh-Dole Act of 1980. This act permitted U.S. universities to own the title to inventions developed with government support, and allowed U.S. government laboratories to grant exclusive licenses to patent. The effect of this is illustrated in figure 5.8 [19].

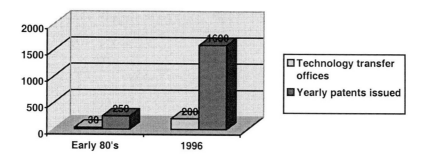

Fig. 5.8: Influence of legislation on patents in U.S. Universities (Bayh-Dole Act)

5.9 CONCLUSION

Distribution channels are often neglected in the nonprofit environment, mainly due to the fact that they create no "visible" feedback. Commercial entities will evaluate marketing channels and can rapidly evaluate the efficiency in terms of sales or turnover.

Distribution of space data from orbit or on ground is a well-known problem and has been studied extensively. The introduction and rapid growth of new communication tools requires a constant updating of the various concepts.

Space activities have led to a considerable spin-off, which requires further development in terms of marketing channels. Not only is it necessary to, more proactively, reach industry to make them aware of potential implementations, but also the general public needs to be reached in view of their support for space activities.

Technology Life Cycles are getting shorter, and therefore the transition from one phase to another has equally to be faster.

Internet techniques have recently proved to be the best distribution channel in this respect, due to the combination of speed, multimedia tools and relatively cheap access. This area is in constant move and new developments are created every day, which are useful also for the space sector. Database marketing will certainly become important the next years and, therefore, also this evolution deserves to be monitored closely.

Contrary to the traditional commercial sectors, the data from space activities are still operating in a legal "quasi-vacuum"; in analogy with the U.S. situation, there is an urgent need to harmonise the Intellectual Property aspects in Europe.

References Chapter 5

1. STERN, L., EL-ANSARY, A. & COUGHLAN, A., *Marketing Channels*, (Prentice-Hall, 5th Ed., New Jersey, 1996).
2. COYLE, J., BARDI, E., & LANGLEY, J., *The Management of Business Logistics*, (West Publ., St. Paul, 1992).
3. KOTLER, P. and ANDREASEN, A., Strategic Marketing for Nonprofit Organizations, (Prentice-Hall, 5th Ed., New Jersey, 1996).
4. ASSAEL, H., *Marketing Management* (Kent Publ., Belmont, 1985).
5. CHRIST, U., FRANK, E, INCOLLINGO, M. & SCHULZ, K.-J., IGS Concept - The Response to COF and ATV Communication Requirements, *Data Systems in Aerospace*, (DASIA 97, Sevilla, May 1997).
6. CHRIST, U., SCHULZ, K.-J., INCOLLIGNO, M. & BERNAL, A., Validating Future Operational Communications Techniques: The ATM Testbed, *ESA Bulletin* No. 92, (ESA, Noordwijk, November 1997).
7. LEUTTGENS, R. and VOLPP, J., Operations Planning for the International Space Station, *ESA Bulletin* 94, (ESA, Noordwijk, May 1998), pp. 57-63.
8. LARSON, W. and PRANKE, L., *Human Spaceflight*, (McGraw-Hill, New York, 1999)
9. DEBUSSCHERE, F., *Marktstrategieën voor hoogtechnologische produkten*, (KUL, Louvain, 1999).
10. ROSENBERG, N., *Perspectives on Technology* (Cambridge University Press, Cambridge, 1976) p.69.
11. BAYUS, B., Are Product Life Cycles really getting shorter? *Journal of Product Innovation Management* 11 (1994) pp. 300-308.
12. POPPER, E. and BUSKIRK, B., Technology Life Cycles in Industrial Markets, *Industrial Marketing Management*, Vol. 21 (1992) pp.23-31.
13. BRISSON, P, BOUGHAROUAT, N. & DOBLAS, F., Technology Transfer and SMEs, *ESA Bulletin*, 101, (ESA, Noordwijk, February 2000), pp.42-47.
14. ESA, *Spin-off Successes*, ESA-BR 152 (ESA, Noordwijk, 1999).
15. PISANO, G. and WHEELWRIGHT, S., High-Tec R&D, *Harvard Business Review*, (September 1995) pp. 94-105.
16. GRIFFIN, A. and HAUSER, J., Integrating R&D and Marketing, *Journal of Product Innovation Management*, Vol.13 (2) (1996) pp.191-215.
17. CAPON, N. and GLAZER, R., Marketing and Technology: A Strategic Coalignment. *Journal of Marketing*, Vol.51 (July 1987) pp. 1-14.
18. MOENAERT, R., SOUDER, W., DE MEYER, A. & DESCHOOLMEESTER, D., R&D-Marketing Integration Mechanisms, Communication Flows and Innovation Success, *Journal of Product Innovation Management*, 11, (1994) pp. 31-45.
19. ISU, *Technology Transfer Design Process*, SSP97 Project, available on WWW.ISUNET.edu/Programs/SSP/
20. BOUTELLIER, R., GASSMANN, O. & VON ZEDTWITZ, M., *Managing Global Innovation* (Springer-Verlag, Berlin, 1999).
21. BUNTE, F., *Product Innovations and barriers to entry*. Ph. D. Thesis (University of Maastricht, 1997).
22. HERROELEN, W. and LAMBRECHT, M., *Innovatie in Produktie*, (Kluwer, Antwerp, 1985) pp.14-18.
23. ROTHWELL, R., Industrial Innovation: Success, Strategy, Trends, in DODGSON, M. and ROTHWELL, R., *The Handbook of Industrial Innovation* (Edward Elgar Publ., Aldershot, 1994).
24. STALK, G., Time – The Next Source of Competitive Advantage, *Harvard Business Review*, 88 (July 1988), pp. 41-51.

25. NAGARAJAN, A. and MITCHELL, W., Evolutionary Diffusion, *Strategic Management Journal*, 19 (May 1990) pp. 1063-1077.
26. TROTT, P., *Innovation Management & New Product Development*, (Financial Times Publ., London, 1998).
27. Lenorovitz, J., 'NASA office to pursue new ideas, technology, *Aviation Week and Space Technology*, November 16 (1992) p.62.
28. BLUME et al., *The Refined Decentralised Concept and Development Support*, Paper presented at COSY 8 (Munich, March 1992).
29. ESA, *The "Radius" Report*, ESA-BR-76 (ESA, Noordwijk, June 1991).
30. ESA, *Impact 2000*, ESA-BR-154 (ESA, Noordwijk, November 1999).
31. Kumar, P., BRISSON, P., WEINWURM, G. & CLARK, J. The Harsh-Environment Initiative Meeting the Challenge of Space-Technology Transfer. *ESA bulletin* 99 (ESA, Noordwijk, September 1999), pp.20 -28.
32. BRISSON, P., *Coming of Age: plastics and space meeting the challenges of mankind*. (ESA, Noordwijk, October 1999).
33. VAN DEN POEL, D., *Response Modelling for Database Marketing using Binary Classification*. Doctoral dissertation nr. 129 (KUL, Louvain, 1999).
34. HARMON, A., Mars Landing Signals Defining Moment for Web Use, *New York Times*, (July 14, 1997)
35. HUDSON, W., *Internet Marketing*, under http://technology.ksc.nasa.gov/ETEAM/SSC/PITCHES/hudson.html (update 21 May 1996).
36. BATES, J., Web Site Gives Researchers A Place to Find Rides to Space. *Space News* (27 March 2000) p.16.
37. LEONG, E., HUANG, X. and STANNERS, P., Comparing the Effectiveness of the Web Site with Traditional Media, *Journal of Advertising Research*, (October 1998) pp.44-50).
38. CLERC, P., in *Workshop on Legal Remote Sensing Issues* (Project 2001, Toulouse, 28 October 1998) p.33.
39. FERRAZANI, M. and THIEBAUT, W., The Legal Protection of Remote-Sensing Satellite Data. *ESA-Bulletin* 76 (ESA, Noordwijk, November 1993) pp.61-63.
40. UN, *Treaty on Principles Governing the Activities of States in the Exploration and Use of Outer Space, including the Moon and other Celestial Bodies*, 19 UST 7570 (UN, London/Moscow/Washington, 10 October 1967).
41. UN, *Convention on International Liability for Damage Caused by Space Objects*, 24 UST 2389 (UN, London/Moscow/Washington, 1 September 1972).
42. ESA, *Intellectual Property Rights and Space Activities in Europe*, ESA SP-1209 (ESA, Noordwijk, 1997).
43. OOSTERLINCK, R., Private Law Concepts in Space Law. In TATSUWA, K., *Legal Aspects of Space Commercialisation*, (CSP, Tokyo, 1992).
44. MALANCZUK, P., The Relevance of International Economic Law and the World Trade Organization (WTO) for Commercial Outer Space Activities, *Proc. Of the Third ECSL Colloquium*, ESA-SP-442, (ESA, Noordwijk, May 1999), pp.305-316.
45. TRIPS, *Trade-Related Aspects of Intellectual Property Rights*. Full text under http://www.wto.org/wto/intellec/intellec.htm, (January 1996).
46. BENETTI, S. and ELIA, C., Co-funded contracts. The New Approach in Support of European Space Competitiveness. *ESA-Bulletin* 99 (ESA, Noordwijk, September 1999) pp. 29-33.
47. KALLENBACH, P., Agency Law and Practice in the Protection of Inventions. *ESA Bulletin*, Nr. 82 (ESA, Noordwijk, 1995) pp. 101-108
48. ESA, *Catalogue of ESA Patents*, ESA SP-1131 (Revision 3), (ESA, Noordwijk, 1999).

Chapter 6

PROMOTION OF SPACE ACTIVITIES

6.1   INTRODUCTION

Wherever appropriate, in the previous chapters the differing applicability of Marketing tools in the nonprofit environment has been underlined. However, those differences are not very expressive on their own, only the implementation thereof underlines the different approaches.

On the other hand, in the case of promotion there is a considerable underlying difference from the essentials onwards. We can recall here the problems mentioned before and put them in the promotion context.

"Marketing is seen as wasting the Public's money"; this statement is specifically the case for advertising campaigns, as for example the six million $ Canadian government campaign to get support for a new constitution in 1980. Specifically for such controversial subjects (contrary to speeding or anti-smoking campaigns), it is easy to understand that the taxpayer, who does not support the specific proposal, will object to the use of their taxmoney to defend such proposal. The essential argument in the Canadian case was summarised as "Good government needs no advertising. It speaks for itself" [1].

"Marketing Activities are seen as Intrusive"; this applies as much to questionnaires as to advertising and promotion in general. Particularly for marketing surveys, there is a widespread fear that Governments might use the information for other purposes.

"Marketing is seen as Manipulative" is the other category where mainly promotion activities are targeted. Promotion campaigns against smoking may, if not properly guided, turn against smokers in general. In the same category fall the anti-AIDS campaigns, which could be interpreted easily as anti-homosexuals campaigns. The Use of threats and fear have to be administered with care and ethical rules are required.

Therefore, the overall promotion or, perhaps better, communication strategy requires a careful step by step planning having, in each step, these aspects in mind. Promotion is the most visible marketing activity of a nonprofit organisation and therefore has to be used with care. On the other hand, promotion is only one of the marketing tools and can be used only in combination with (some of) the other tools, never as a self-standing and independent approach (which is unfortunately sometimes the case).

The above certainly explains why the text in this chapter is less applicable to business-to-business space marketing and business-to-consumer marketing of space data (such as navigation systems). In these cases the traditional promotion tools are fully applicable, whereby the specific visionary aspects of space certainly can be used in the background. Therefore, most of the remarks are more related to the promotional activities of space Agencies and public space authorities.

6.2 PROMOTION VERSUS COMMUNICATION

The term Promotion is used here in its "classical" form as a Marketing Mix tool. In classical terms, we further distinguish between:

- Advertising
- Personal selling
- Sales promotion, and
- Public Relations.

In terms of nonprofit organisations and our specific subject in particular, they are defined by Kotler and Andreasen [2] p.516 as:

Advertising: Any paid or unpaid form of non-personal presentation and promotion of an event by an identified sponsor through a formal communications medium.

Personal selling: Oral presentation of information about an event in a conversation with one or more prospective target audience members for the purpose of securing support

(Sales) promotion: Short-term incentives to encourage support for a project or an event.

Public relations: Non-personal stimulation of publicity for an event by securing the reporting of commercially significant news about the event in a published medium or on radio, television or another medium that is not paid for by the sponsor.

So, public relations are only one of the elements, even in the classical promotion definition. Therefore it is misleading that in many nonprofit organisations the related department is still called "Public Relations".

Kotler and Andreasen [2] p.478 note that for the nonprofit environment Communication is the most important factor, therefore they strongly propose the use of this term over Public Relations and Promotion for the nonprofit sector. Compared to the more classical approach, communication also covers elements such as:

Sponsorship marketing: promoting the interest of an organisation by associating to an event or other product (corporate image)

Point-of-purchase communications: encompasses displays, posters, signs, but also packaging and brand naming.

In view of the criticism on "paid" communications in the nonprofit sector, it is of importance to classify these instruments as per table 6.1

	PAID	FREE
MASS	Advertising Promotions Point of purchase	Publicity Corporate Image Sponsorship
PERSONAL	Personal Selling	"Word of Mouth"

Table 6.1: Characteristics of Communication Tools

The rapidly changing information society evidently also has an influence on these "vested" methods. Shimp [3] points out that there are four major effects:

1. **Reduced Faith in Mass Media Advertising.**

Different communication means have competitive and sometimes even more effective results than the traditional mass media advertising. This was not the case in the past, mainly due to technical constraints, but presently other communication methods should be carefully considered before "automatically" assuming that mass advertising is the appropriate solution.

2. **Increased Reliance on Highly Targeted Communication Methods.**

Many alternative media now allow selective communication with target groups. One of the most prospective areas is so-called "database marketing", using selective computer technology (one simply has to think of "cookies" of which each Internet user is familiar with). The implementation of these techniques will eventually allow the sending of tailored information to specific user groups, in this way, enhancing the efficiency and, at the same time, reducing the costs.

3. **Greater Demands imposed on Marketing Communications Suppliers.**

The classical advertising agencies and PR firms are gradually converting to cover a wider span of services, at the request of their clients. Therefore they will consequently and possibly automatically offer more aspects than the client may initially have had in mind.

4. **Increased Efforts to assess Communications' Return on Investment.**

With modern computerised tools, it is much easier to correlate the results of an action with its effect (for example: the number of people who click on a banner can measure the result of such an Internet banner).

Shimp [3] p.12 relates these changes to **Integrated Marketing Communications** (IMC), defined as:

> *IMC is the process of developing and implementing various forms of persuasive communication programmes with customers and prospects over time.*

Five elements are of specific importance in IMC:

1. Affect the Behaviour.

People first have to be made aware of a product before they will change their purchasing behaviour. An example often quoted is a 1994 survey where it was found out that 43% of the sampled consumers thought that free-range chicken were superior to conventional chicken, whereas only 6% bought them. Assuming that people appreciate your product does not automatically make them buy it.

2. Use all forms of Contact.

Any mean should be explored to bring a uniform message and a clearly recognisable brand label (TV, Internet but also car stickers, T-shirts). Essential here is the uniform appearance of for example a logo (also uniform colour schemes) or a message (Example the Bayer-aspirin relationship).

3. Start with the Customer or Prospect.

The approach is also labelled "outside-in", as it starts with the customer to determine those communication methods that will best serve the customer's information needs and motivations. In other words, as we will describe further in the tools, knowledge of the target audiences and the key messages for those specific audiences are a prerequisite to set up a promotion campaign.

4. Achieve Synergy.

This is mainly the preparatory phase in-house, before a campaign is brought to the outside world. All of the communication elements must speak with one voice. Coordination of the messages is essential to achieving a strong and unified brand image.

5. Build Relationships

Essential is the link between a brand and a customer, which eventually may even lead to loyalty. Constant mailing and e.g. frequent-flyer programmes are some examples from the commercial markets. Presence at conferences and symposia are more applicable instruments in the space environment.

Even if they are not popular in each environment, campaigns such as "Intel Inside" (with the accompanying characteristic music) and "Windows 95" are examples of an effective IMC approach. Every computer buyer will be interested if there is an "Intel Inside" and if "Windows" is installed, not based upon their knowledge, but based upon the multimedia marketing campaigns that they have been constantly confronted with. As an example, when Windows 95 was launched on 24.8.1995, a 200 million $ campaign was used to support the introduction, using a variety of means [4], p.575:

- In New York, the Empire State Building was lit up in the colours of the logo
- In Toronto, the CN Tower bore a Windows 95 banner
- In London, an editorial supplement was placed in the "Times of London" newspaper
- World-wide commemorative T-shirts were distributed
- A theme song ("Start Me Up" from the Rolling Stones) was launched.

The various communication effects can also be hierarchically organised, using the so-called "a-b-e" model [5], which stands for

- Attribute focus (what is typical for the specific product?). This approach is recommended for expert target audiences and intangible services.
- Benefit focus (what does the buyer want?), recommended in case of "hard-to-imitate" benefit.
- Emotion focus (what does the buyer feel before or after the benefit?), recommended in case of "easy-to-imitate" benefit.

In accordance with this evaluation, space communication shall focus on the "a" and "b" levels, avoiding emotional elements.

6.3 EUROPEAN SPACE PROMOTION

A number of elements were perceived within ESA as sensitive with respect to promotion activities:

- The risk of criticism from Industry or other Agencies; in the case of extensive promotion, this is not only relatively quickly criticised as wasting taxpayer's money, but also as an element of competition (even if all organisations would benefit from a better space image in Europe).
- "Good news is no news". Press has in general a certain negative attitude towards space activities and knows that articles describing a mishap are

well read. It certainly demands skilled staff to reply to press questions trying to "provoke" certain statements in interviews.
- A basic choice to promote "Space", and not ESA, was the logical consequence of these considerations.

However, this turned not only against ESA but also against the whole European space sector once geopolitical changes and the consequent ending of the "Space Race" made space activities less popular. As result of a workshop on this topic [6], the following basic action plan was laid down:

- Identify and prioritise target groups
- Develop different messages for various target groups and update message to fit wider media trends
- Increase quality and decrease quantity of information material for the respective target groups:
  - faster, cleaner
  - more attractive presentation to enhance understanding
- Messages from the space community need to be clear
- Orient space information material so that it is relevant to:
  - Government priorities: employment, economic growth,
  - daily life, by relating experimental results to wider goals
- Update and adapt messages and information regularly
- Remind audiences that the original economical/industrial justification for public space expenditure is still valid.

In 1998 a survey was made; based upon 8350 "weighted" respondents from an ESA Member States telephone inquiry made by a professional company this showed that:

- 42 % of the respondents has an interest in space exploration
- 63 % of the general public think an European Space Programme is important
- only 12 % of the general public spontaneously knows ESA (on average, but the percentage widely varies from country to country)
- but… 54 % spontaneously know NASA

An external company was hired to examine the problem in more detail; for the general public four propositions were selected, namely the influence of ESA on

- Delivering scientific knowledge
- Delivering economic value
- Delivering benefits for daily life
- Delivering protection to the Earth and mankind.

The general public saw none of these propositions in 1999 as a proper justification of a strong European organisation. On the other hand, two other criteria in general were brought forward:

- A more ambitious scope and focus on wider and long-term benefits for mankind
- An organisation that is not in the shadow of NASA but has its own ambitions.

Also in this survey, the level of spontaneous "Brand Awareness" of ESA was low (between 10 and 40%, depending on the target group)

As far as the two aspects mentioned are concerned, they deserve some deeper examination. The "ability to expand the frontiers of mankind" remains obviously the main motive for general public support in space activities. We can also illustrate this differently. If one examines the selection of the 100 best advertisements, as presented in [7], one notices that approximately 10% make use of space and space exploration to "sell" their message. This is certainly not limited to technological firms, but for example also for Prudential Life Insurance ads. A similar effect can be noted in European magazines, where Audi refers to spaceframe technology used in its cars and even the Dresdner Bank's advertisements are based upon a futuristic Space Station. This aspect will be elaborated in the next chapter: Space exploration as part of mankind's cultural progress.

The comparison with NASA explains to a large extent the lack of impact of previous campaigns. There is a considerably different "corporate image" created by NASA, especially after the Pathfinder and "Life on Mars" publicity. The interactive way NASA uses the Internet can be illustrated by figure 6.1; the "Cool site for Kids" is designed to communicate with the target audience very effectively and is written in an audience-oriented language.
As far as the NASA effect is concerned, we could illustrate this (in a slightly exaggerated way) as per table 6.2

CHARACTERISTIC	NASA	ESA
Image	Leader in space exploration	Cooperation with industry and other Agencies
Attitude	Military, secretive	Peaceful, sharing
Spirit	Exciting, dynamic	Trustworthy
Tools	Competitive	Cooperative
Focus	Benefit of the U.S.	Benefit of Europe and global community

Table 6.2: Comparison NASA vs. ESA image

Fig. 6.1: NASA's "Cool Site for Kids" [26]

On spontaneous reactions, the general public associates NASA with (in ranked order, top five):

- Space flights
- Moon landing
- American
- Space Shuttle
- Leader in space exploration

The same public associates ESA with:

- European
- Satellite
- Space flights
- Ariane
- Cooperation with other space agencies.

Clearly, in analogy with table 6.2, these are "softer" achievements compared with the more spectacular NASA ones. This correlates with the more discrete ESA style, compared to the rather "flamboyant" NASA style, specifically promoting inter alia the MARS missions, which clearly appealed to the general public.

6.4 DESIGN OF A NEW COMMUNICATION STRATEGY

The results of previous investigations did not pass unnoticed. After a competition, a contract was awarded to the company Saatchi & Saatchi (a major communications firm) in 1998, mainly to study the improvement of the ESA image and come up with an action plan proposal. Soon the effort focused on three aspects:

- Produce a corporate brand identity which corresponds to a role for ESA on which there is agreement from all audiences
- Define which audiences are the priority for communications
- Devise a structure for the effective management of communication – external and internal.

This coincides with a general approach for an effective communication strategy, as proposed by Kotler and Andreasen [2] p.483, which is summarised in 6 steps:

Setting of communication objectives

The objectives of the communication programme need to be well defined in advance. From the list suggested in [2] we could adapt this for European public space communication as:

- Awareness of space programmes
- Changing negative beliefs of space expenditure
- Changing perceptions about the European space organisations
- Influencing decision makers.

Generating possible messages

Messages can be collected in a different number of ways:

- A customer oriented approach, by talking to the customers and analysing their desires about the messages
- Brainstorming meetings with key personnel within the organisation.
- Use of a formal, deductive framework. In our case specifically the use of rational messages (for example the search for the origin of life) seem the most appropriate tool with which to establish such framework.

Overcoming selective attention

Some research suggests that we are confronted (or shall we say bombarded?) with some 1400 messages daily. Therefore one has to select the style, emotional value, wording and form of a message in a way that it catches the target audience. However, for nonprofit organisations this leads very fast to conflicts with ethical norms. The use of fear in nonprofit marketing is a typical subject which has led to a lot of controversy in this respect.

Overcoming perceptual distortion

Individuals have a substantial background of experiences, prejudices, needs, wants and fears that can strongly affect what they "see" or "hear" in the message. This can work to the advantage of communicators, because messages can be "bundled" into a few words even using associative colours; however, it can also lead to "back-firing". There exists, indeed, a specific danger for culturally different perceptions of the same event or message as will be described later.

As an illustrative example, we can compare the ads for Levi's jeans in different cultures [4], which were carefully chosen to associate with local perceptions:

- In the U.S., emphasis is put on outdoor life, showing healthy young individuals in outdoor settings (perception of the "American Dream")
- In Japan, James Dean is used as a symbol, relating to the "fashionable" and rebellion aspect
- In the U.K., a male in underwear in a Laundromat is shown, using the humorous perception.

Choosing a medium

The message to be transmitted needs the appropriate choice of a medium or different media. The medium can be personal or impersonal and will be perceived by the receiver as advocate or independent. Examples are given in table 6.3

	PERSONAL	IMPERSONAL
ADVOCATE	A speaker (e.g. astronaut) An organisation spokesman	Brochure Advertisement
INDEPENDENT	Newscaster Known academic	Government report Specialised newspaper

Table 6.3: Alternative Media

Evaluating and selecting messages

From the different messages, the best alternatives have to be selected in terms of

- **Desirability**: the message should say something on why the project or event is being promoted.
- **Exclusiveness** : the message has to suggest a distinction with alternatives
- **Believability**: the statement must be provable and defendable.

There is no doubt that individually each of the steps have to be taken into account on a case-by-case basis. However, it is felt that in the case of communicating space and space projects the main emphasis should be put on three major aspects:

- Which message do we want to convey?
- Which target audience do we choose?
- Which means do we use?

These three aspects will be elaborated in the next chapter, as they also coincide with the present communication re-evaluation process ongoing in ESA.

6.5 A EUROPEAN SPACE MESSAGE

Pan-European surveys indicate that European space activities have a number of "strong" elements, which can be used as positive messages. They include mainly four elements:

- ESA is a good demonstration of multi-cultural cooperation
- ESA has a tradition of openness
- ESA has proven to work at high quality standards
- ESA keeps an overview of global issues (versus country specific objectives).

These elements, generally accepted, could be used as background for messages. The major problem seems to be more the lack of vision that people find in NASA and obviously is not so identifiable in Europe. Therefore it is probably of more importance to develop a number of attractive messages.

Pragmatism must be constantly monitored in this context. With the funds available in Europe for basic space activities, one has to be pragmatic. One could never envisage e.g. to plead for a manned Mars project, because the costs required would exceed by far the available funding. Therefore, from the pragmatic point of view, affordable projects have to be envisaged which will, however, fill a "gap" in the space activities "market", without duplicating for example NASA efforts (where for this case the word "market" bears no relation to commercial aspects). Examples in the past are the Ariane success story (launching heavy satellites at a lower cost than the – manned – Shuttle) and a number of novel scientific missions, such as Giotto, that performed a first close encounter with a comet. Global objectives for the 21st Century are presented in [8] as:

- Improve the quality of life
- Ensure human survival
- Increase global wealth
- Further exploration of the universe
- Search for our origins and destiny
- Exploration of space resources.

Some key recommendations are

- Maintaining a World-competitive European launcher
- Take the initiative for an International Lunar Programme
- Keep the lead in climate and environment monitoring
- Promote a programme for natural and technological disaster detection and monitoring
- Develop technologies for future telecommunication and navigation systems.

The report concludes:

> ***Thus, Europe's choice for tomorrow is not whether space activities – be they commercial, public-service, military or exploratory – will continue to expand or not. Europe's choice is whether to play its part in that irreversible expansion in a way commensurate with its cultural heritage, political stature and economic potential.***

Within this conclusion and these themes we find a number of messages which could boost the prestige, if properly communicated, for the general European public (similar to the Ariane prestige). Based upon the results of these activities, the mandate of the Long-term Space Policy Committee (LSPC) was extended and a more detailed report was produced, under the chairmanship of Peter Creola. In the final report [9] twenty potential European issues for the next few decades are grouped in four categories, as per table 6.4

Without any doubt each of these points, which were the result of a broad European expert panel, have important merits; however, they will appeal to different audiences in different ways. Industry will encourage a number of the points, as they are the essential prerequisites with which to tackle other, more visible ones (such as topics 2, 3, 6, 7 and 8), and so will the policymakers.

THE CHALLENGE OF INDEPENDENCE	THE CHALLENGE OF PLANETARY MANAGEMENT
1. Search for Earth-like planets 2. Cheaper Access to Space 3. Innovative Space Station Utilisation 4. Future Navigation System 5. European Space System for Security and Peacekeeping 6. Creation of a European Telecom Regulatory Body 7. Small Business Innovation Initiative 8. Micro-miniaturisation Technology Initiative	9. Space Monitoring of Compliance with Environmental Regulations. 10. Disaster Warning from Space 11. Space Weather 12. Space Debris 13. Threat of Cosmic Collision
THE CHALLENGE BEYOND	REACHING OUT
14. Telepresence Demonstration Project 15. European Lunar Initiative 16. Space Energy and Resources 17. Weather Modification from Space	18. European Space Education Programme 19. Public-Awareness Initiative 20. European Space Policy Institute

Table 6.4: European 2000+ Space Action Plan

For the general public, some of the points may be too technical and indirect to "market" as a vision, but certainly other ones are appealing to a large audience. This is specifically the case with those effects, which have an impact on every day's life. In this category we can certainly put:

- **Security and Peacekeeping**

Recent conflicts have demonstrated Europe's dependence on non-European surveillance satellites and their resulting information. The Peacekeeping effect, by early detection of for example concentration of troops has a proven a priori effect (giving diplomacy a chance to intervene in an early stage). As some conflicts are now closer to Europe, it is probable that Europeans may be sympathetic to a move in that direction.

- **Spaceborne Early Warning systems**

Specifically in the field of earthquakes and volcanic explosions, many lives could be saved by early warnings. Whereas in the past this was more evident outside Europe (see table 6.5), recent earthquakes such as those in Turkey have shown that Europe is vulnerable to this as well.

YEAR	LOCATION	MAGNITUDE	FATALITIES
1999	Izmit, Turkey	7.4	15,000+
1993	South India	6.3	30,000
1990	Northern Iran	7.7	40,000+
1988	Northwest Armenia	7.0	55,000
1978	Northeast Iran	7.8	15,000
1976	Tangshan, China	8.0	255,000
1976	Guatemala	7.5	23,000
1970	Northern Peru	7.8	66,000

Table 6.5: Major earthquakes in the course of the satellite era (source [10])

As far as volcanic explosions are concerned, major European fatalities have not occurred in recent years (see table 6.6) but one should not forget that four European volcanoes are classified as potentially dangerous (date of the last eruption in brackets), namely:

- Etna (Italy, 1998)
- Stromboli (Italy, 1998)
- Santorini (Greece, 1950)
- Vesuvius (Italy, 1944)

YEAR	VOLCANO	LOCATION	FATALITIES
1991	Pinatubo	Philippines	800
1986	Lake Nyos	Cameroon	1,700
1985	Nevado del Ruiz	Colombia	23,000
1982	El Chicon	Mexico	1,880

Table 6.6: Major volcanic eruptions in the course of the satellite era (source [10])

CHAPTER 6 : SPACE AND PROMOTION   203

(Photo : CNES/CLS/ARGOS)

Fig. 6.2 : ARGOS system used for surveillance of volcanic eruptions (Volcano Kelut)

- **Impact of a large Asteroid on Earth**

The Tunguska asteroid which struck Siberia in 1908, had an impact equivalent to an explosion of some 50 Megatons; meaning that it would have been able to destroy completely a metropolis. Impacts of such asteroids became more "visible" to the larger public through the images of the 20 impacts of the Shoemaker-Levy Comet fragments collision with Jupiter between 16-22 July 1998.

Statistically, impacts of this size are estimated to happen once every 300 years on Earth, whereas 20 Kiloton events happen every year. Although there are still some discussions ongoing, it is now accepted that the impact of a large asteroid 65 million years ago was responsible for a global world catastrophe, for which a similar event in the future could even mean the end of human life on Earth. It goes without saying that this knowledge increasingly worries the general public and any remedial step in that direction would be welcomed.

Projects which will require longer periods for the development, such as a European presence on the moon or the experimental use of Space energy will certainly equally boost the image of European space activities. The carriers to achieve this goal are undoubtedly contained in the "Outreach" group of proposals, such as Public Awareness and Educational initiatives. In this context we have to return to the "image" aspect.

Bromley [12] emphasises the differences of corporate image between the private and the public sector. The reputation of organisations in a free-market economy is largely based on the attributes thought to reflect its competitiveness (size, efficiency, and profitability...). The reputation of public (service) companies is more based upon performance and expected standards. Therefore the author concludes that [12] p.180:

The common factor that accounts for the relationships between the variables that contribute to reputation would be the overall capacity of an organisation to make itself visible through the impact its performance has on those sections of society where its reputation matters.

ESA has adopted in 2000 a "Shape and Share" message, expressed as follows:

With ESA, Europe shapes and shares space for people, companies and the scientific community

In this context **Shape** refers to the commitment to build and to meet the raising the challenges of space, whereas **Share** is about disseminating discoveries among the international community.

6.6 THE SPACE AUDIENCE IN EUROPE

Globally we can distinguish four "audiences" in Europe for ESA:

- Policy makers
- Space partners (other agencies and industry)
- Media
- General Public.

The previous chapter has clearly demonstrated the different interest for each of these groups, so the key messages may be different for each audience as well. Moreover, there are different interactions and communication channels. There is a direct contact, even a reporting line, to the policy makers. From this point of view this audience is without any doubt the most influential one. The influence of policy can be illustrated with following historical example:

In the first half of the 15th Century, China had a fleet of 3500 ocean going ships, with trade routes to India, Japan, Arabia and Africa. No doubt the discovery of America would have been a next step (and global world policy would have developed from then onwards rather differently) if policy makers had not interfered. As they found ocean-going fleets too expensive and questioned further expansion (since China already had the biggest fleet), they stopped production of ocean going ships. Within only decades, the West overtook the technological lead and China rapidly fell into isolation.

Evidently, commercial space applications are living partially independent from this (even if regulations and treaties have a considerable impact), but exploratory space activities still strongly depend on government funds. In a democratic system policy makers are influenced by the general public, but also by the space partners (for example through industry in view of employment or capacity problems) and by the media.

However, there is a difference in the time horizon. Whereas communication with policy makers has a direct influence, this has a more delaying factor through the media. Communication with the general public, on the other hand, is more a long-term investment. Many space organisations have realised the importance of involving the younger generation and actively pursuing educational programmes, e.g. the CNES education programme (accessible from the CNES homepage, www.cnes.fr) or "NASA for Kids" (under www.nasa.gov/kids.html). Therefore we can categorise the impact for the different target audiences as per table 6.7.

TARGET GROUP	CONTACT	EFFECT
Policy Makers	Direct	Immediate
Partners	Direct	Mid-term
Media	Partially direct	Short-term
General Public	Mainly indirect	Long-term

Table 6.7 Effect of communication with target groups

In classical marketing terms the indirect contact with the general public is probably surprising and may need some clarification. A number of elements influence the decision to reach the general public indirectly through the media, mainly:

- **Financial considerations**

The public is a very large and differentiated group; one could distinguish four categories, each of which would require a different approach:
- Space interested, ESA aware
- Space interested, ESA unaware
- Space unaware but interested
- "Great Unwashed".

In order to reach a positive influence by direct approaches, such as advertisements, would require a very expensive campaign.

- **Linguistic problems**

Not everybody speaks English, and in some countries this number is even less than generally assumed. When the commercial satellite systems started entering the market, a Gallup survey was made in 1985 [12] p.477. If we concentrate on the knowledge of English and French (the official ESA languages) in European households, we come to the following extract of the survey as represented in table 6.8.

COUNTRY	PERCENTAGE OF ADULTS SPEAKING ENGLISH	PERCENTAGE OF ADULTS SPEAKING FRENCH
Norway	80 %	10 %
Denmark	51 %	5 %
Netherlands	50 %	16 %
Germany	30 %	12 %
France	26 %	100 %
Belgium	26 %	71 %
Switzerland	26 %	55 %
Italy	13 %	27 %

Table 6. 8 English/French spoken in European households

Although we may assume that on average the figures have increased since then, it is clear that one only reaches a portion of the general public when using only for example English and French brochures. Consequently, the costs would increase considerably if the uninformed target group has to be reached in a "pan-European" effort.

- **Reluctance to use paid publicity.**

Any nonprofit organisation is careful when using public funds for promotion, because this often leads to criticism from the "donor" organisations. A successful approach to overcome this problem is shown in figure 6.3. The figure shows a picture which is extracted from publicity for perfume whereby the ESA logo is discretely incorporated (a good example of the sponsorship marketing approach).

(By courtesy of Parfums Givenchy)

Fig. 6.3: Example of use of unpaid publicity for corporate image improvement

6.7 THE CHOICE OF MEDIA

If we start with a global framework, ESA internal assessments correlate the different media and different categories of general public as per table 6.9.

TARGET GROUP:	BROADCAST	POPULAR TV	NATIONAL PRESS	POPULAR PRESS
Space interested, ESA aware	XX		XX	
Space interested, ESA unaware	XX	X	XX	X
Space unaware but interested	X	XX	X	XX
"Great Unwashed"		XX		X

XX: prime medium, X: interesting medium

Table 6.9: Reach of target groups per medium

Whereas this approach is very generic, tuning is needed to account for the cultural different patterns in media consultation. If we take the expenditure as a basis, but re-express it in relative terms, then we can note the considerable differences per country from a 1995 survey [4] p.545

COUNTRY	TV	PRINTED	RADIO	CINEMA	TRANSIT	TOTAL
France	29	51	7	1	12	100
Germany	16	76	4	1	3	100
U.K.	32	62	2	1	3	100
U.S.	35	52	11	0	2	100
Japan	40	39	5	0	16	100

Table 6.10: Relative Publicity expenditure per medium

We can deduce from table 6.10 that:

- Print media are still very important in the U.K. and, especially Germany
- Japan, with densely populated cities, is interesting for outdoor/transit publicity such as posters. The same is valid for France (specifically Paris).

- In particular commercial TV stations are changing the picture considerably. One can safely assume that more recent data will reflect an increase of TV publicity in Germany, France and the U.K.

Finally, if we now add on top of this some cross-cultural differences in presenting the content of the message, such as [13] p.342:

- In Germany : Rational, descriptive, informative
- In France : Innovative, modern, attention-getting
- In the U.K. : Subtle, understated, humorous
- In the U.S. : Emotional, lifestyle, glamorous,

then it soon becomes clear that in Europe one single communication approach, in one single language is doomed to fail. This has led to an awareness in ESA of the need to install "country-desks" for communication. Besides these differences we should also not ignore that various media reach a different public and have different effects, as we can note from table 6.11 [15] p.171.

EFFECT	TV	RADIO	MAGAZINE	NEWSPAPER
Total population reach	+++	+	0	+
Adult reach	0	+	++	+
Cost per 1000 persons reached	+	+++	++	+
Emotional stimulation	+++	0	0	-
Ability to catch attention	+++	-	+++	+
Brand name	+++	+	0	0
Prestige of the medium	0	0	+++	++

+++ = Very strong
++ = Strong
+   = Good
0   = fair
-   = poor

Table 6.11: Characteristics of mass-media

This would suggest for space activities:

- The use of TV to catch the attention of a large public for popular messages
- The use of magazines for more technical/cognitive subjects
- The use of TV for messages which are designed to create reactions (for example the asteroid collision scenario).

The advertisement aspect has not been touched in this context; this requires some explanation. Research in nonprofit organisations teaches us that in general [15]:

- Advertising is not a very efficient means of communication for nonprofit organisations
- Paid advertisements are too expensive.

If we consider that this research was executed in Canada, with a more established openness for PSAs (Public Service Advertisements), there is no doubt extreme care to be taken in considering paid advertisements.

This opinion is largely endorsed by Rotschild [17] p.728 who examined the effect in a number of specific cases, as summarised in table 6.12.

CASE:	SITUATION INVOLVEMENT	COST/BENEFIT RATIO	PRE-EXISTING DEMAND	SUCCESS OF COMMUNICATION
Military enlistment	Very Complex	Very good for certain segment	Very specific	Possible
55 mph speed limit	Low	Poor	Virtually none	Low likelihood
Antilitter	Low	Poor	Low	Short-run impact possible
Voting	High to Low	Favourable (for voting)	Moderate	Short-run impact likely

Table 6.12: Non-business communication case studies

Rotschild concludes that a campaign must satisfy a number of prerequisites to be accepted, such as

- In the public interest
- Simple to advertise
- Non-partisan, non-political
- Timely.

It is doubtful that space activity advertisements would coincide with these criteria as specifically the subject is too complex to present in a simple form, there is a strong partisan/political component and the space activities are rather time-independent.

The cost factor is important in view of the ratio between cost and the public reached and taking into account the response. Theoretically we could express this as:

$$c = \frac{T}{N} \cdot E$$

Where:
c = Cost of communication per person reached
T = Total cost of the communication effort
N = Number of people reached
E = Effectiveness of reach.

The effectiveness factor E is evidently varying strongly with the medium used (e.g. 0.3 % for an email banner) but this can be compensated by the higher reach.

For Hi-Tech products and projects, research seems to indicate that the differences are not as great as for consumer products. Urban [17] for example, examined the effect of convincing medical doctors to use a new analysing technique, mainly to test the efficiency of multimedia techniques (such as Computer Aided Instructions). As per table 6.13, he came to the conclusion that the technique used did not give considerable differences in results when convincing the target audience, contrary to differences in consumer products such as e.g. automobile sales.

Technique Used	Convincement Factor
Brochure	4.3
Advertising	4.3
Sales Presentation	5.1
Article in Journal	5.7
Simulated presentation	4.9

Table 6.13: Response factor for Hi-Tech Medical equipment

As far as general public outreach is concerned, it might come as a surprise to find an excellently written English booklet from as early as 1983 on "Economic Uses of Space Technology" edited in Russia [18]. Besides the obvious political motive of the booklet it is remarkable to see the visionary attitude of the authors, at a time when in Europe very limited efforts were made to propagate this area. The leaflet describes the direct effects of the different types of satellites in layman terms and links this very efficiently to such themes as "Television in Every Home", "Linking Countries and Continents" and "International Search and Rescue Services". The cover page of the booklet is shown in figure 6.4

Fig. 6.4: Cover of Russian booklet (1983) [18]

6.8 A SPECIFIC PROBLEM: THE IMPORTANCE OF NAMES

All authors involved in cross-cultural problems stress the importance and the care that must be taken when choosing a name which needs to be valid and acceptable in a multilingual environment. Gesteland [19] p.116 illustrates this with the example of Italian toilet tissue which had unexpected problems on the English market, although the brandname "SOFFASS" did very well in Italy...

Ricks has been concentrating on the topic of "Big Business and Marketing Blunders". From one of his books [20], we can summarise some problems encountered with technical names as per table 6.14.

BRAND NAME	LANGUAGE PROBLEM	REASON
Chevrolet "Nova"	Spanish	No va = it does not go
Car "Randan"	Japanese	Randan = idiot
AMC's "Matador"	Puerto Rico	Matador = killer
Ford's "Fiera"	Spanish	Fiera = ugly old woman
"Bardak" machinery	Russian	Bardak = brothel, chaos

Table 6.14: Examples of language problems in product names

The lesson from this is clear: names must be chosen carefully. Good examples of this are "Kodak" and "Exxon"; both names were developed with extensive computer-assisted linguistic research and very lengthy discussions with linguistical specialists on pronouncibility in foreign languages. International space projects will have to learn from this. Specifically when technical abbreviations are innocently used, one should not forget that e.g. in the ISS world they will be confronted with some 15 different languages.

The different perception of model, brand or project names is studied in [23]. Any name has to be chosen on the basis of its appearance in different languages, as illustrated by the well-known name of the French group "Creusot-Loire", whereby:

- The hard "CR" is unusual in English
- The French "EU" is virtually unpronounceable for Americans
- The "S" must be pronounced "Z" in French (unknown in English)
- The "T" is mute in French, contrary to other languages

- The "OI" is a typical French "diphthong", unknown in English
- The "E" at the end is mute.

Obviously, such a name is difficult to memorise for non-French and the chances for misspelling are very considerable. For new names, a number of devices are suggested in [21] p.294, grouped as:

- Phonetic devices (how does the name sound?). Preference is suggested for alliterations and consonance, as well as rhythm (e.g. Cocoon, Vizir, Tic-Tac). Another recommendation is to start with initial plosives such as b, d, k and t (e.g. Bic, Dash)
- Orthographic devices (is the name easy to write?). Avoid unusual writing which are typical for one language (apostrophes, special writing in French and Scandinavian languages)
- Morphological devices (creating an association with the item), such as Trans-Rapid, Tipp-ex
- Semantic devices (referring to culture based interpretations or associations). Examples in this category are Hawaiian Punch, Bounty, Midas, and Ajax.

It has to be noted here that in Russian space activities more emphasis has been put on this element in the past. Let us take the example of MIR; the word stands in Russian language as well for "Peace" as for "World", creating a philosophical association. Moreover it is easy from a phonetic point of view and causes very few orthographic problems in any language.

6.9 ETHICAL ASPECTS OF SPACE COMMUNICATIONS

Whereas ethics can be defined as

The principles of moral conduct governing an individual or group

business ethics comprises moral principles and standards that guide behaviour in the world of business. Applying the "opportunities for ethical misconduct" to the space environment we have to take into account the following factors:

- **Intangibility of certain services**

It is virtually impossible for the general public to verify or control certain results, such as the use of space originated methods in other sectors, or the costs of a launch. Hence, the opinion is a priori based upon the information accuracy of the provider himself.

- **Technical Complexity**

The results of scientific missions have to be "translated" by the provider in order to be understood by the general public. Here also, the provider needs an ethical morale.

- **Lapse between Project Initiation and Outcome**

The famous saying "when the answer is given the question is forgotten" is highly applicable to cases with a 5-7 years project lifetime. It is virtually impossible to relate both statements a posteriori.

Also important in this context is the difference in being informative or being persuasive. Information can contain a number of "hidden" persuaders. Providing information on osteoporosis, just to give one example, will certainly provoke indirectly support to space activities from a large part of the audience, in particular those suffering from the disease. From the viewpoint of behavioural science [22], three active messages can be given, namely:

- **MODIFY** a current opinion ("choose mine. It suits you best")
- **STOP** performing an activity ("don't do this. It is harmful to you, to others or to society in general")
- **START** a new activity ("do this. Yourself, others or society need it").

Here also, within the boundary conditions of avoiding criticisms, an essential choice has to be made by space organisations. It looks like the third message ("start supporting space, because ...") is the only acceptable one to be applied for this specific environment. The consequences of such ethical aspects in space activities can be illustrated with the following example:

There are a number of convincing arguments, such as the impact of an asteroid collision, to which the general public is susceptible. This effect has been amplified by a number of spectacular movies based upon such a scenario, such as "Armageddon". The positive side of this is an awareness that space activities could contribute actively to reducing the risk, in the first instance by the early detection of collision risks.

A lot of research has been carried out on the use of fear in particular, and emotions in general using Public Service Advertisements (PSAs). PSAs are widely used, especially in the U.S. as an important part of social marketing (at a rate of not less than 4 billion $ yearly). A generally accepted model is that a threat in such advertisement (be it warnings against cancer, AIDS, drugs or child-abuse) shall not be of danger to the message recipient himself but rather to someone else. The objective of this threat is to motivate people to help others in danger; thus the appeal is indirect. This corresponds to a general theory of emotion called the "Lazarus Model".

In skeletal form the Model proposes that the appraisal processes of internal and situational conditions lead to emotional responses and these in turn, induce coping activities; schematically [23]:

Appraisals → Emotional responses → Coping

Applied to a "threat" in an advertisement, a positive effect is reached if coping results in empathy and a desire to help or contribute, thus:

Negative emotion (threat) → Empathetic responses → Decision to help.

The authors used two different test ads and demonstrated that emotional anti Child-Abuse ads were more effective than rational ads in generating empathy for victims and the decision to help. From this theory and associated experiments we can deduce that there is no basic problem to using threats provided that the psychological response to the threat is positive. The differentiation and a tool to make such differentiation are described in [24], whereby the authors conclude:

> *The purpose of assessing communications effectiveness is not to find an optimal level of fear, but to determine the optimal type of threat that a given target audience segment will act upon.*

Schoenbachler et al. [25] made a survey on the messages to reach young drug users. Also they came to the conclusion that anti-drug messages involving a physical threat had rather a boomerang effect because the target group are precisely the sensation seeking youngsters. Therefore they argue for social disapproval messages such as "drugs can ruin your reputation" which, in their test, scored much higher than the physical threat approach (note that they found as a side result that 40% of the schoolchildren had used drugs in the last 30 days; a figure that obviously even astonished the researchers).

The reason why the drug example is used is the remarkable social threat approach of NASA on the previously mentioned www.nasa.gov/kid page [26]. The straight message is that "you can't do really neat things like being an astronaut or a NASA scientist if you are on drugs. The cool thing to do is to Explore Space ... Not Drugs". The social threat in this case being that drugs may ruin the prospects for a challenging career or youngster's dream. Statements by astronauts, as for example Buzz Aldrin reinforce the message:

Space Travel satisfies the insatiable curiosity of all mankind. Distractions like drug and alcohol hinder our ability to pursue the answers.

The "threat" approach, if properly managed, could be employed for the threat of "townkillers" (large celestial bodies, which reach the Earth causing an impact large enough to erase a big town and that are statistically arriving on average every 300 years). The message to the general public should avoid a "doom" emphasis but appeal to supporting space activities by the prospect of earlier detection and eventually collision avoidance. Here again we have to stress the problem of "information reliability".

It is a documented fact that on the morning of 30 June 1908 a catastrophe took place in the "Podkamenoj Tunguski" ("The Stony Tunguska"). The area is so vast and inhabited that it took years before information on this event reached the scientific community. In fact it was some 20 years before a first scientific expedition took place which, of course, does not simplify the task of reconstructing the events. Many of the evidential accounts from the few witnesses provided contradictory information at that time and no substantial material has been found. Although the "meteorite-theory" is the most probable one, comprising a core of crystals (methane, carbon dioxide, and ammonia) which led to an enormous explosive chemical reaction, other versions include tectonic event and even the explosion of a UFO, as summarised in [27].

This example brings us to the ethical aspects of marketing in general and communications in particular. Many approaches try to present a theoretical framework but have problems to "translate" this as realistic guidelines. Unfortunately, in most cases ethical aspects are ignored until a real problem occurs which, in its turn, then generally results in overreaction and overscrutiny. The most pragmatic advise can be summarised as follows:

- Compile an "internal code of ethical conduct"
- In case of any doubts, apply external evaluation boards.

The involvement of external advisers is considered by Kotler and Andreasen [2] p.566 because of three reasons:

1. There is a careful balance to be maintained between the interest of the organisation and individual freedom. The nonprofit public advocate, often in his conviction to fight for the right case must weigh each proposal against the general value of preserving individual choice and behaviour.

2. One must be careful not to "demonise" the other side (an example of this is the "health vigilantism" against smokers). Being fully convinced of a certain case is not a valid argument to consider non-believers as bad.

3. One must avoid paternalism. Other people may have other preferences; one does not always have to share their opinion but, on the other hand, they can have justified reasons for their specific case.

The risk is bigger when specialists who are fully convinced of their opinion are confronted with such situations. Inappropriate reactions can even lead to antipathy for the case in general (example: TV round table discussions). Porras and Weinberg [28] develop a pragmatic approach for evaluating the effects of possible communication actions which goes along following steps:

1. Establish a panel.
It is important that this panel is broadly composed in terms of nationalities ("culture-free messages"), social status and so on, to represent as close as possible the target audience.

2. List all possible critical points.
This could be done in a brainstorming session where for each proposal the question is put "can anybody be hurt by this message". A typical aspect could for example be: does the message "sound" as if the service will be free of charge?

3. Assign probabilities.
For each case the likelihood of a problem shall be estimated preferably by individual figures and a consensed figure afterwards. At this stage, there is a

chance that ethical conflicts can already be eliminated based upon predetermined go/no-go criteria. If this is not the case, the following and more subjective steps are proposed.

4. Sequence the possibilities.
Based upon consentient probabilities from the previous step, the expectation is that alternatives can be found that cause problems with much lower probabilities (which is of course only the case for multiple proposals)

5. Formalise Utility Assignments.
This is the last possibility of coming to a solution within the panel. The evaluation is rather subjective and should basically answer the question "is it worthwhile pursuing this proposal considering the potential risk?". One could assume that this last step will not be necessary in a "sensitive" environment, such as it is the case in most nonprofit situations. Rarely it will be worth taking a risk in view of the possible and traditionally over-dimensioned reactions.

The above described procedure has the merit to be pragmatic but does not solve the essential question on which norms to be applied. Therefore, a number of authors suggest to apply "Hypernorms", whereby the norms of the target group shall prevail [29]. A "Kantian" approach to this follows a number of questions, namely:

1. Does the intended action violate the law? (legal test)
2. Is the action contrary to widely accepted moral obligations? (duties test)
3. Does the action violate any special obligation inherent to the organisation? (special obligations test)
4. Is the intent harmful? (motives test)
5. Is there a likelihood that people or organisations will be damaged by the action? (consequences test)
6. Is there a satisfactory alternative without such problems? (utilitarian test)
7. Is there any possible infringement on property rights or privacy rights? (rights test)
8. Does the action leave one group less well off? (justice test)

Specifically in conjunction with point 3, any approach will need a specific "ethical code" tailored towards each type of organisation.

6.10 CONCLUSION

The present situation is unfortunately not very satisfactory for Europe in terms of public Promotion for Space activities and a lot will have to be done to remedy this. It is clear that many parameters have to be taken into consideration when working out a communications plan, which do not facilitate the task.

An Integrated Marketing Communications plan is a potential solution for broadening the traditional PR concept. A good communications plan and a proactive approach will pay off eventually. We can best illustrate this with the NASA example: despite two high-profile failures during 1999 for the Mars missions, respondents of a 2000 survey gave NASA an 82% favourable rating, 7% higher than a comparable 1999 survey [30].

The use of different media has to be considered in terms of (cost) efficiency and must take into account the changing communication means such as the Internet. Paid advertising must be avoided because the drawbacks are not in balance with the limited outcome.

An essential element, which is often lacking is a "code of ethics" for which only guidelines can be given but to which each individual organisation needs adhere, in accordance with the organisation-specific rules. In Europe, cultural differences and linguistic problems aggravate the situation with this.

The elements which certainly will appeal in Europe, unfortunately contain an element of threat (security, earthquakes, impact of an asteroid). However, it has been demonstrated that this element, if properly managed, is useful in generating an empathetic reaction. Key messages for the next decennial in Europe should focus on more independence in space activities, planetary management and broader outreach ("Shape and Share").

In order not to create a too dark picture, at least one positive result shall be mentioned here. In Germany a survey [31] was made of children between 6 and 14 years old. Whereas in 1994 only 1 % of the boys wanted to become an astronaut, this was already 2.2% by 1999. This placed "astronauts" in the top ten, although it was still behind such professions as football player (1st), policeman (2nd) and even truck driver (9th).

REFERENCES CHAPTER 6

1. MALCOLM, A., Ottawa Runs Into Protests Over Its Huge Advertising Costs. *The New York Times* (1 November 1980) p.10.
2. KOTLER, P. and ANDREASEN, A., *Strategic Management for NonProfit Organizations*, (Prentice-Hall, 5th Ed., New Jersey, 1996).
3. SHIMP, T., *Advertising, Promotion, and Supplemental Aspects of Integrated Marketing Communications*, (Dryden, 4th Ed., Orlando, 1997).
4. JOHANSSON, J., *Global Marketing*, (Irwin, Chicago, 1997).
5. ROSSITER, J. and PERCY, L., *Advertising Communications and Promotion Management*, (McGraw-Hill, New York, 1997).
6. ESA: *Approaches in Communicating Space Applications to Society*, ESA-SP-384, (ESA, Noordwijk, May 1995)
7. POPPE, F., *100 New Greatest Corporate Ads*, (Wiley and Sons, Chichester, 1993).
8. ESA, *Rendez-Vous with the New Millennium*, ESA SP-1187 (ESA, Noordwijk, 1995)
9. ESA, *Investing in Space, The Challenge for Europe*, ESA SP-2000 (ESA, Noordwijk, 1999).
10. NEW YORK TIMES, *World Almanac 1999*, (World Almanac Books, New Jersey, 1999)
11. BROMLEY, D., *Reputation, Image and Impression Management*, (Wiley and Sons, Chichester, 1993)
12. JEANNET, J.-P. and HENNESSEY, H., *Global Marketing Strategies*, (Houghton-Mifflin, 2nd Ed., Boston, 1992) p.477
13. NORGAN, S., *Marketing Management. A European Perspective.* (Addison-Wesley, Wokingham, 1994).
14. KOTLER. P. and ROBERTO, E., *Social Marketing* (The Free Press, 1989).
15. MARCHAND, J. and LAVOI, S., Non-profit Organizations' Practices and Perceptions of Advertising: Implications for Advertisers, *Journal of Advertising Research*, (July, August 1998) pp. 33-40.
16. ROTSCHILD, M., *Advertising*, (Heath, Lexington, 1987).
17. URBAN, G. et al., Information Acceleration, *Journal of Marketing Research*, (February 1997) pp.143-153.
18. NOVOSTI PRESS, *Economic Uses of Space Technology*, (Novosti Press Publ., Moscow, 1983).
19. GESTELAND, R., *Cross-cultural Business Behavior*. (Copenhagen Business School Press, 1999).
20. RICKS, D., *Big Business Blunders, Mistakes in Multinational Marketing*, (Irwin, Homewood, 1983).
21. USUNIER, J.C., *Marketing across Cultures*, (Prentice-Hall, 2nd ed., 1996).
22. FENNELL, G., Persuasion: Marketing and behavioral science in Business and Nonbusiness contexts. In BELK, R. *Advances in Nonprofit Marketing Vol.1* (JAI Press, Greenwich, 1985) pp. 95-160.
23. BAGOZZI, R. and MOORE, D., Public Service Advertisments: Emotions and Empathy Guide Prosocial Behavior, *Journal of Marketing*, Vol.58 (January 1994), pp. 56-70.
24. LATOUR, M. and ROTFELD, H., There are Threats and (Maybe) Fear-Caused Arousal, *Journal of Advertising*, Vol. XXVI (3), (Fall 1997) pp.45-58.
25. SCHOENBACHLER, D., RIDNOUR, R. & HUMES, K., Sensation Seeking and the Effectiveness of Physical and Social Threat Drug Prevention PSA Messages, *Journal of Nonprofit & Public Sector Marketing,* Vol.4, nr. 1 / 2 (1996), pp. 51-73.
26. NASA, *NASA Web Site for Kids*, www.nasa.gov/kids (update 19 September 1999).

27. JÄHN, S., *Die Tunguskakatastrophe aus der Sicht eines Kosmonauten*. Paper presented at the International Tunguska Meteorite Conference (Schoenfels, Germany, 13 December 1998).
28. PORRAS, J. and WEINBERG, C., A Framework for Analyzing the Ethics of Marketing Interventions, in MOKWA & PERMUT, *Government Marketing* (Praeger, New York, 1981) pp. 356-375.
29. DUNFEE, T., SMITH, C. & ROSS, W., Social Contracts and Marketing Ethics. *Journal of Marketing,* 63 (July 1999) pp. 14-32.
30. BERGER, B., NASA, Exploration Score High in Public Opinion Poll, *Space News,* (April 17, 2000) p.14.
31. FOCUS, Wenn ich eimal gross bin…, *Focus* 35, 30 August 1999

Chapter 7

SPACE AND PHILOSOPHY

7.1 INTRODUCTION

It is not the intention to add another P here for the sake of being inventive or to distinguish from other authors. However, the feeling remains that the most important reason why mankind is involved in space exploration is not a purely economic endeavour but it is an integral part of a higher motivation. When confronted with water, people found means to cross the water by building boats. When confronted with long distances they invented transportation methods and finally they conquered airspace by developing "Heavier-than-Air-Flying-Machines".

It is inherent to human nature to conquer new frontiers, in order to make progress, even if this raises scepticism and criticism. Some enthusiasts have baptised the conquest of space as the "Final Frontier", but certainly there will be other frontiers to be conquered after this one, even if they are less tangible now. The idea originally stems from the famous book written by Wernher Von Braun and called "Space Frontier" [1] where it was firmly stated that the conquering of space is the main challenge in our present times.

From the moment that people became aware of the existence of other planets and celestial objects, it is evident that one way or another they started to dream on how to reach them. We should not forget that without any additional means such objects could be seen with the naked eye and that is not the case for remote areas, which were known only by verbal stories and could only be pictured in one's imagination. This is much less the case with for example the Moon and the Sun which could be seen every day. Therefore it is understandable that human imagination has invented methods to reach these objects, from Daedalus and Icarus to Jules Verne.

224 SPACE MARKETING

Fig. 7.1 The "Space Frontier" idea (1963)

The dream of space travel always existed and remained alive throughout the centuries. When Neil Armstrong put his foot on the moon in 1969, there were an unprecedented number of people following this event on TV irrespective of time differences. It is this aspect that still appeals most to people. Space exploration needs to be done in an economical and affordable way and the previously described "classical" 4 Ps are an integral part of general acceptance from the public. But more than these aspects, the "Frontier" element is a paramount one, and therefore it cannot be imbedded under the umbrella of the previous ones.

Moreover, there is a second philosophical dimension which only now begins to appear. Space and celestial bodies belong to nobody; from this point they would be able to give mankind a new chance to "do things better". Legal instruments and Treaties will have to be developed, hopefully having the bad experience of the last centuries in memory. In order to develop space activities, people have to closely work together but driven by economic realities. People now must also realise that they have to learn to work together and forget past prejudices. The first social "testbed" in this respect is the International Space Station which will be described in next chapter. Here, two generations from different (political) cultures, who were taught as children that the others were the enemies have physically to cooperate. This leads to a gradual process of mutual understanding, a process we could summarise as "Different is not wrong".

Still, a lot of criticism can be found about space activities in general, irrespective of the omni-present enthusiasm. In order to assess this contradiction, one has to look into the history of science and innovation in order to see that such attitude is probably inherent in human nature itself.

7.2 TRADITIONAL RESISTANCE TOWARDS CHANGES

More than 500 years ago, the "Council of Wise Men of Salamanca", after three years of deliberations about the proposal of Christopher Columbus concluded that:

> *There can be no justification for your Majesty's support for a project based on extremely weak foundations and plainly, to anyone who knows about such things, impossible to achieve*

The same type of comments has appeared, in a different form but with the same content, on any major occasion or change in technology, such as [2]

On train travelling:

> *Train travel at high speeds is not possible because passengers, unable to breathe, would die of asphyxia (D. Lardner, M.D., 1859)*

On alternating current:

> *Fooling around with alternating current is just a waste of time. Nobody will use it, ever. It's too dangerous, it could kill a man as quick as a bolt of lightning. Direct current is safe.*
> *(T. Edison, inventor, 1880)*

On airplanes:

I have not the smallest molecule of faith in aerial navigation other than ballooning. (Lord Kelvin, physicist, 1870)

Flight by machines heavier-than-air is impractical and insignificant, if not utterly impossible. (S. Newcomb, physicist, 1902)

On nuclear power:

The energy produced by the breaking down of the atom is a very poor kind of thing. Anyone who expects a source of power from the transformation of these atoms is talking moonshine. (E. Rutherford, physicist, 1930)

And of course, on space:

That Prof. Goddard does not know the relation of action to reaction, and the need to have something better than a vacuum against which to react. (...) We hope that the professor from Clark College is only professing to be ignorant of elementary physics if he thinks a rocket can work in a vacuum. (New York Times, 1920)

Space Travel is utter bilge. (Sir R. Wooley, astronomer, 1956)

Similar arguments are now heard about manned Spaceflight, missions to Mars and so on. It is evidently difficult to forecast results over longer periods, but luckily this has not stopped pioneers in the past continuing to explore beyond known frontiers and to travel faster and higher. Probably the eminent scientist Niels Bohr was correct when he stated:

Prediction is very difficult, especially about the future.

In previous quotes the timing and professions of the originator are mentioned on purpose. Amazingly one notes that in general we are dealing with professionals in the subject area and the fact that many quotes were made very shortly before demonstration of the technology in subject. Peters and Waterman [3] refer to this phenomena as the "paradigm shift", a concept which was introduced in a broader perspective by Kuhn [4] and which may explain these quotes as follows. Scientists in any field and in any time period possess a set of shared beliefs about the world. For that point in

time, the set constitutes the dominant paradigm. "Normal Science" proceeds under this set of common beliefs and carries out experiments within these boundary conditions. Each time a considerable change in opinion emerges, this acts as a "quantum leap" in progress, and this is the paradigm shift Kuhn was the first to refer to. Some examples of this are:
- Copernicus and Kepler, attacking the Ptolemaic view of the Earth as a centre of the universe
- Einstein's relativity theory, in contrast to the paradigms of Newtonian physics
- Acceptance of tectonic plates in geology.

This approach could explain very well the present resistance, also from certain scientists, against a full development of space activities, as was similarly the case every time mankind was on the verge of conquering a next frontier. It is difficult now to defend manned Spaceflight with tangible arguments, just as it was 40 years ago to defend the interests of unmanned Spaceflight. But critics in the past have again and again proven to be wrong and there is not one reason in the world to believe that they are right this time.

7.3 THE SLOW PROGRESS OF SPACE DEVELOPMENT

It would be too easy to "excuse" the relatively slow progress in space activities over the last few decades by the previous explanation. Space exploration started very fast, nurtured by political motives and made a number of spectacular successes as we can describe using table 7.1. If we chronologically place the major milestones of space development and just arbitrarily awarding them one "step" each; we come to following table:

MILESTONE	DATE	EVENT
1	1957	First satellite launched (Sputnik)
2	1961	First man in space (Y. Gagarin)
3	1965	First commercial satellite (EARLY BIRD)
4	1969	First man on the moon (N. Armstrong)
5	1986	First part of MIR launched
6	1998	First part of ISS launched
7	?	First manned mission to Mars

Table 7.1: Milestones in Space Exploration

The earlier steps were part of the "Space Race", whereby specifically the U.S. and the (then) Soviet Union were fiercely competing to put the first man in space and to be the first in the "Race to the Moon". As mentioned earlier, considerable budgets were made available to support this development and, as we can see from table 7.1, approximately every four years a success was recorded.

After the first man stepped on the moon, suddenly the pressure fell away and further progress fell back rapidly (the post-Apollo effect). It is for example typical to see how the full Soviet Moon Programme, even if it was in an excellent shape (and probably only failed to beat the American one with a timeframe of some weeks) was completely stopped. Moonlanders, which were virtually ready to take off, were mothballed and never employed afterwards. We have mentioned the 4-year step sequence in table 7.1. Although there is no real scientific value or reliability in the following, if just for the sake of comparison we would extrapolate the nearly linear progress from the first few milestones, we can note the considerable time lapse in progress as per figure 7.2,

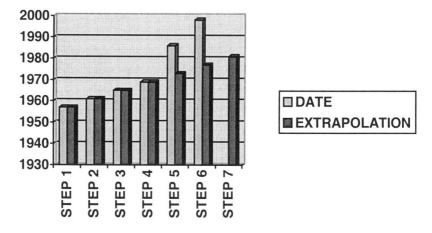

Fig. 7.2: Extrapolated versus real progress in space milestones

Using this as a comparison tool, we note how strongly progress is backtracking after the Apollo era. We can also note that the extrapolated dates are in the same order as forecasts made in the 1960's, when for example, Von Braun estimated that step 7 in figure 7.1 would take place in 1986 [1] p.199.

Technological progress cannot be blamed either for this lack of progress. In Koelle [5] we can find an interesting table comparing some key technologies and their evolution in the timeframe we considered before in table 7.2.

TECHNOLOGICAL PARAMETER	1960	1975	1990
Launch capacity (in kg)	12	12,000	120,000
Number of transponders in orbit	15	15,000	1.44×10^6
On Board Memory capacity	0.5 Mb	20 Gb	8000 Gb
On Board energy capacity (kWh/kg)	0.04	80	800
Active switch capacity per cm^3	0.3	10,000	0.3×10^8
Orbit calculation precision (in m)	1000	50	0.1

Table 7.2: Evolution of technological performances over time

Although the parameters from table 7.2 are relatively random, they clearly show the exponential growth of technological capacities. Furthermore, they refer to a number of key elements, essential for any space project. Therefore, technological progress cannot be considered as a contribution to a slower space evolution. An even more pessimistic (or fatalistic?) theory is presented in [6]. Based upon the main milestones as presented in table 7.3 one could deduce that main human milestones only occur in "cycles" of 60 years.

DATE	PERSON	EVENT
1850	D. Livingstone	Exploration of Africa
1909	R. Peary	North Pole Expedition
1969	N. Armstrong	First man on the moon
2030	?	First man on Mars?

Table 7.3: Major milestones in human exploration

If this "theory" is correct, we are currently in a "deep" and a space enthusiasm revival is still a while away. Luckily in the past most forecasting theories have proven to be too static and were largely overtaken by the dynamism of mankind.

A more realistic observation is that space activities are entering a certain vicious cycle. Due to the geopolitical changes it was more difficult to maintain public interest in space. As a result of this the public space expenditure gradually reduced. On top of this came a long economically difficult period, with the combined effect that most important space budgets have been gradually reduced. In trying to keep up with the changed budgets, remedial actions such as the "Faster, cheaper, better" were invoked in order to try and maintain to the outside world a constant progressive image. This has unfortunately led to a number of failures which again are reflected in less public support for space activities and, consequently, less political support.

It is not easy to break this vicious circle, and even more difficult to move it in another direction. However, the right foundations are still present: people believe in space activities and are enthusiastic about it deep down. It is a major challenge to space agencies and industry to "re-market" this enthusiasm.

7.4 CHANGES AND INNOVATIONS IN THE SPACE SECTOR

The standstill in progress has also led to studies and assessments on the microeconomic scale. In general, a number of changes were strongly felt in the U.S. in the period 1960-1980 in the consumption market which can best be described by following quote [7] p.37:

> *We thought they could never catch up, but they tried harder, and here we are – to paraphrase an old American slogan, a Sony in every house and two Toyotas in every garage.*

If we take the example of the automobile industry, which was a considerable part of America's industrial culture, they saw indeed that the sales of Japanese cars rose from 0.25% to 22% on the American market in the period 1960-1980. This represented a hundredfold increase over 20 years with a devastating effect on the American automobile industry. When productivity figures were compared, the output per labour hour rose 3.4% in U.S. for the period 1970-1975 and only 1.6 % in the period 1975-1980 which can be compared to respectively 6.7% and 7.9% in Japan. This led to a number of attempts at bringing in Japanese know-how particularly in the area of cars and consumer electronics.

It was in that period that Kanter [7] performed some studies and came to the conclusion that the problem was laying within the American industry itself. In her famous book of 1984 "The Change Masters", she urged industry to become once more open to innovative approaches and to change rapidly from what she introduced as "segmentalistic structures" to "integrative structures". Perhaps being faced with very little competition after the Second World War, industry had become almost bureaucratic, by following rules as provocatively described in [7] p.101 and summarised in table 7.4.

> **Rules for Stifling Innovation**
>
> 1. Regard any new idea from below with suspicion – because it's new, and because it's from below.
> 2. Insist that people who need your approval to act first go through several other levels of management to get their signatures.
> 3. Ask departments or individuals to challenge and criticise each other's proposals. (That saves you the job of deciding; you just pick out the survivor.
> 4. Express your criticisms freely, and withhold your praise. (That keeps people on their toes.) Let them know they can be fired at any time.
> 5. Treat identification of problems as signs of failure, to discourage people from letting you know when something in their area is not working.
> 6. Control everything carefully. Make sure people count anything that can be counted, frequently.
> 7. Make decisions to reorganise or change policies in secret, and spring them on people unexpectedly. (That also keeps people on their toes.)
> 8. Make sure that requests for information are fully justified, and make sure that it is not given out of to managers freely. (You don't want data to fall in the wrong hands.)
> 9. Assign to lower-level managers, in the name of delegation and participation, responsibility for figuring out how to cut back, lay off, move people around, or otherwise implement threatening decisions you have made. And get them to do it quickly.
> 10. And above all, never forget that you, the higher-ups, already know everything important about this business.

Table 7.4: Segmentalistic "rules"

The rationale for "segmentarism" is the assumption that problems can be solved when they are carved into pieces and the pieces assigned to specialists who work in isolation. Companies with segmentalistic cultures also showed a segmentalistic structure: clearly separated departments, clear distinction between management layers and strong task differentiation between Headquarters and field offices. As a result of this, the structure discourages innovation. Mechanisms are meant to keep the organisation on its planned course but, at the same time, they inhibit change and innovation. Functions that are compartmentalised create barriers for communication and the exchange of ideas, from which innovation should emerge. Based upon a broad research project, Kanter was able to demonstrate that there is indeed a good correlation between the company culture and structure and the number of innovative products brought to the market.

Her global conclusion is that "to meet the Japanese challenge" it is not necessary to adopt Japanese management techniques but rather the awareness that [7] p.64

The problem before us is not to invent more tools, but to use the ones we have.

The rationale behind this is to restore the American entrepreneurial spirit that was the key to success for the country before. Kanter refers to this urgent change as the "American Corporate Renaissance".

Recently, by looking to the past, Micklethwait and Woolridge [8] have come also to a similar conclusion in assessing the "Art of Japanese Management". The Toyota factories, a "company city" two hours out of Tokyo, were indeed based upon months of studying the American automobile industry by Kiichiro Toyoda and Taiichi Ohno (the Toyota founders) in the 1950's. They mainly added to their observations that compared to Japanese organisations there was too much "muda" – a Japanese term that encompasses wasted effort, wasted material and wasted time.

The reason why this worked in Japan lays deeper in the cultural dimension of Japanese people, embodied in the fundaments of Zen philosophy. The doctrines of continuous improvement ("kaizen") and of shared decision-making ("ringi") are part of Japanese culture and do not need to be imposed. Even in their relatively segmentalistic and autocratic organisations, this runs as a "matrix" through the organisations, hence the success of for example quality circles.

Trying to "transplant" these techniques to another culture, such as proposed in the management best-sellers of the 1980's "The Art of Japanese Management" [9] and "The Mind of the Strategist" [10] are doomed to fail as the cultural foundation on which they are based is not present.

Authors like Kanter [7] and Peters [3] have therefore argued for "fighting back with old, proven tools", for which the basis used to exist (such as entrepreneurship). If we notice what happened economically after 1980 and especially in the 1990's with the booming American industry and declining Asian markets, it seems that their message has not passed unnoticed.

There is a specific reason why this notion is elaborated here in such detail. There is a remarkable resemblance between the observations made in the competitive sector of consumer goods in the middle of the 1980's and the present state of space activities. Innovation appears to be hindered by organisations that are too "segmentarian" and therefore we may draw valid conclusions from the previous case (whereby the "Japanese car manufacturers" in previous case are probably the new "End-to-end space commercial entities" in the space sector).

This resemblance can be illustrated by another aspect noted by previous authors. Kanter [7] p.99 points out that a considerable number of innovations are made by what she calls "Lone Rangers". She describes them as organisational loyalists acting on their values to remedy what they see as less-than-optimum situations for a company and a job they care about. Such a person often induces changes and innovation on their own initiative by often breaking rules or bypassing superiors.

An example quoted is the development of the IBM 360 system which was to be stopped on top management orders. Researchers in San Jose, despite several reminders, believed in the low-power concept and continued to develop it even against explicit orders (some of them were even subject to disciplinary action). Ultimately, IBM management had to recognise the concept and incorporated it in the product line where it turned out to be a 5 billion $ winning product.

This approach fits with another concept which was introduced around that same period by Potter, namely the "Corporate Ronin" [11]. The word Ronin originally refers to free samurai who had left the service of their feudal lords in Japan. This took place mainly in the 17th Century when the Japanese feudal system collapsed. In view of their education and skills, they played a considerable role in the Japanese change from feudalism to industrialism.

The concept became widely known from the book written in 1645 by one famous Ronin called Miyamoto Musashi and that became very popular in the 1980's under the title "The Book of Five Rings" [12]. Potter sees their contribution to the innovation process as follows [11] p.192:

The acts of a myriad of individuals drive the innovative organisation. There would be no innovation without someone somewhere deciding to shape and push an idea until it takes usable form as a new product, management system, or work method. Corporate Ronin create new possibilities for organisational action by testing limits and by pushing and directing the innovation process. They have an intellectual elasticity or flexibility that enables them to come up with realistic responses to changing situations. (…) Ronin bring many benefits to the corporation. As a result of many nonlinear moves and a wide range of experience and skill, for example, they tend to see problems in a larger perspective and understand all the important operations of the company. Having moved laterally within the company, Ronin have allies in different departments and at different levels, enabling them to build coalitions to get things done. Ronin are able to envision new possibilities, and when organisational structures and protocol block actualising their vision, they are inclined to bend the rules.

A survey of employment in the space sector, as reported in chapter 2, has revealed that it is exactly this innovative aspect that attracts modern "Ronins" when applying for jobs in the sector. It can only be hoped that the space environment will continue to provide them with a sufficiently receptive ground with which to express their aspirations. (We should not make the parallel here that each Samurai had two choices when leaving his lord; either to "go Ronin", or to commit "Sepukko" (ritual disembowelment)…).

Whereas the previous authors have looked into this aspect from a sociological angle, there is a philosophical support to this view also. Based upon Heidegger's philosophy of human nature, Steiner [13] suggests that the unconventional individuals are central factors to innovation success, rather than conventional scientists or engineers. However, recently the effect that space activities will attract highly talented people seems rather to be decreasing. NASA claims [14] to have difficulties in finding and attracting new recruits especially at post-doctoral level. Besides the job security aspect (linked to budgetary cuts) it is also felt that the booming Internet companies obviously have a higher appeal these days for modern "Ronins". The effect is clearly sector-wide; the age pyramid is rapidly shifting in the U.S. aerospace sector as reported in [15] and shown in figure 7.3.

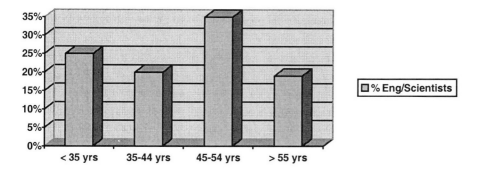

Fig. 7.3: Current age distribution in U.S. Aerospace sector (status: 2000)

This means that 54% of the current workforce at engineer and scientist level in U.S. Aerospace industry is over 45 years old, with 33% eligible for pension in the next 5 years. Also the aerospace industry cannot compete with the salaries, bonuses and stock options offered by Internet and Venture Capital firms. Inventive recruitment schemes such as offering free college tuition to high-schoolers and presentations at middle-school levels by space engineering veterans are now being initiated to reach the next generation of potential engineering recruits.

As an additional and more global remark, the same Kanter published an article on the interaction between the nonprofit and the profit sector in the change process [16]. Based upon research it is demonstrated that a number of successful programmes (such as Bell's Project Explore, IBM's Reinventing Education and Mariott's Pathways to Independence) have provided an excellent testbed in the nonprofit environment and also of benefit for the traditional companies in the field of innovation. In a slightly different context, we can relate these findings to the previously mentioned "Dual-Use" concept. She concludes that this might be

> *(…) a new way for companies to approach the social sector: not as an object of charity but as an opportunity for learning and business development, supported by R&D and operating funds rather than philanthropy.*

7.5 THE UNDERLYING PHILOSOPHICAL DIMENSION

In order to come to some essential answers, DLR created a project composed of space scientists and philosophers to group the technical, scientific and ethical aspects of manned space flight. This programme called "Saphir" came to the conclusion that [17]: "Manned Spaceflight has a cultural dimension because it changes the picture of mankind and his place in the Universe (...) exploration of the Universe is a logic further conquest of frontiers, typical of the universality of man and in line with his sense of responsibility".

The conclusions have further been elaborated in [18], wherein the authors (a scientist and a philosopher) stress the "Trans-utilitarian rationale" whereby space activities shall not exclusively be judged in terms of costs and gains. Leading roles in space exploration and enhancement of a nation's prestige have been a number of political motives which have been used in the past. But also promotion of international cooperation and the pioneering aspects should not be ignored, not even the colonisation of celestial bodies and advancement of human civilisation which can contribute to a new basic outlook on life. Specifically on space pioneering they conclude:

> *The historical argument addresses the special U.S.-American tradition of pioneering into unknown territory, "to see, what's behind the next hill". A more general view holds that the urge to transcend boundaries, physical and mental ones, is a deeply rooted longing of the human species. Leaving the sphere of the familiar and venturing into the unknown, often perilous, areas serves to extend human potential in its broader sense, comprising intellectual, scientific, technological and economic possibilities.*
>
> *(...) The Space Frontier, for some the ultimate barrier for mankind to break, represents today a great challenge for human ingenuity, requiring not only large resources, but at the same time setting free great energies.*

Note that observations of this nature can also be imbedded in the relatively new discipline called "historical anthropology". An attempt to do this can be found in Kerner [19] who defends space exploration as a "natural" continuation of mankind's evolutionary process.

Previously, Zabusky [20], an anthropologist of Princeton University, has studied cooperation in European space activities from an ethnographic point of view. Based upon interviews with scientists and engineers of ESA, the author came to the conclusion that cooperation was driven by the higher motivation of people in the space sector which is describes as "Sacred Cooperation and the Dreams of Modernity". As an overall evaluation [20] p.219:

> *In working together on space science missions, participants who in so many ways can be identified as elites (as the keepers and purveyors of "formal" knowledge, see Foucault) themselves experience chaotic uncertainties and perceive sacred meanings in the world (...). In the ideals of truth, unity, and impartiality, these participants identify a world resonant with meaning; in their continual striving to reach these, they give voice to sacred aspirations, to dreams of communion and epiphanies of collective creativity in work.*

Activities in this context have been continued in Europe under the impulse of P. Sahm, by organising a bi-yearly Workshop called "Man and Cosmos". The first one took place in Nordeney (Germany) in October 1996. In the introduction [21], p.7, the multidisciplinary approach is justified as follows.

> *In this context the question is certainly justified whether human kind should restrict themselves to their earthly habitat or venture into the space frontier.(...) For answering and solving such questions and problems a cross-fertilisation type thinking is required, a polyhistoric, interdisciplinary approach. On the one side fears must be overcome, on the other uprooting has to be avoided – in any case, the dialogue between the technical "realisor" and the personality with a liberal arts background, i.e. the historian, the philosopher or even the theologian, must be cultivated.*

The "fears" and concerns may not be so evident at a first glance but we should not forget certain ideas such as to genetically transform people by using biogenetics and yielding innovative types of metabolism that, for example, would make man less sensitive against the dangers from cosmic radiation. Moreover, if one thinks about the proposals of F.J. Tippler [22] who even suggests to simply carry genetic material in bioreactors on space journeys (instead of human beings), then it becomes evident that, in analogy with e.g. genetic cloning, indeed a broader forum of opinion is needed.

This brings us to the deeper philosophical background of space exploration. It is not a coincidence that the previously mentioned Workshop refers to the important work of Teilhard de Chardin [23]. (Note that the German title of this book is "Der Mensch im Kosmos"). Originally, according to the "Monadologie" of Leibniz, there was a philosophical opinion that the world is mankind's natural home [24].

> *As it has been in the mind of God to select one World out of the infinite number of possible ones, this is the ultimate proof for the choice of God that we belong to this World, more than to any other one.*

It was the acceptance of the anthropic principle, first in the philosophical world, later also in the theological one, which made room for another vision.

We can distinguish between the "soft" anthropic principle, which we can translate as "in view of the fact that there are living creatures in the universe, the universe must have a number of properties which allow the existence of such living creatures" and the "hard" version. The "hard" anthropic principle goes much further and can be described as "the Universe must have been established in such a way that once it produces living creatures". This formulation does not only opens the possibility for an acceptance of life on our planet, but also on other planets and hence on space exploration. Moreover it may lead to interpretations of a "More-World-Hypothesis".

Acceptance of the anthropic ideas have in any case led to philosophical broadening in planetary exploration and in space exploration in particular as well as in theological circles. One could also illustrate the philosophical dimension with the wording of the 1967 Outer Space Treaty [25]. Specifically Art.5 stipulates:

> *States Parties to the Treaty shall regard astronauts as envoys of mankind in outer space and shall render to them all possible assistance in the event of accident, distress, or emergency landing on the territory of another state Party or on the high seas.*

Even in this legal text astronauts are considered as "envoys of mankind in outer space", highlighting their broader exploratory role. Moreover, such cooperation is not only formal as a commitment. During the problems with Apollo-13, the full Russian space recovery system (including a fleet of special ships) was offered to assist NASA and re-positioned in order to

CHAPTER 7 : SPACE AND PHILOSOPHY   239

assist in case of an unforeseen landing event. This illustrates – in the middle of the Cold War period – how space activities can link people in view of common objectives.

One has to add a more pragmatic aspect to this. Even if there are still some discussions taking place between scientists (if this would not be the case, they would be no scientists...), there is a broader consensus that the planet Earth is vulnerable to collisions with a larger celestial body, such as a big asteroid. It seems still not to be clear to people in general that what once happened to the dinosaurs some 65 million years ago could also happen to them. Moreover, we can express this in probabilities and sit back for another few million years but statistically, it could also happen next year. The 5 km diameter Toutakis asteroid passes the Earth every 4 years at only some 4 million distance (a "near-miss" situation in astronomic dimensions), any minimal deviation from this orbit could become dangerous. Mankind has no means to react against such event even if noticed well in advance. Only some theoretical evaluations are performed, such as the model developed in [26]. This model concludes that asteroids larger than 1 km in diameter can be deflected from collision by application of a single surface burst on a reasonable time scale, whereas asteroids with diameters of less than 0.5 km will have to be dispersed.

Still, these theoretical approaches are not followed up by practical tests and early detection programmes. NASA has problems to finance the 75 million $ Deep Impact programme, whereas, on the other hand, the Armageddon movie had a 250 million $ budget allocation [27]. Neglecting these facts could lead to the extermination of the whole mankind in one go that with the exception of some fatalistic beliefs, cannot be in the spirit of any philosopher.

This aspect on its own should already be enough to promote colonisation of other planets in order to "spread the risk" as Carl Sagan expressed in a Keynote Address of a NASA symposium. In his opinion, there is a 1-in-1,000 chance that an object of 1.5 km in diameter could collide with Earth within the next 100 years, releasing an energy likely to kill more than a billion people. Consequently, he concludes [28]:

> **A significant human presence in the inner solar system beyond the Earth is mandated (…) it is safer for the human species if we're on many worlds than if we're on only one…**

Unfortunately, obviously no politicians or political party have thought about putting this in an election manifest up to now. Many people have expressed in the past the philosophical aspect of space exploration. We cannot avoid quoting here one of the most famous ones; the Russian "father of space activities", Konstantin Tsiolkovsky, wrote already in 1903:

The earth is the cradle of mankind, but one cannot remain in the cradle forever

He was one of the many visionary people, in history, who believed in the philosophical dimension of progress. Mankind tries to expand its frontiers, even against continuous criticism. In Europe, this was expressed by Reimar Lüst, when Director General of ESA as follows [29]:

The peaceful exploration of space began, for Europe, as a dream for some farsighted scientists. Today it has far outstripped their intentions, but it is our duty, and our privilege, to keep the spirit of their ambitions alive.

At the same time, this also illustrates the close relation between space science and philosophy. Tsiolkovsky, without much outside communication, worked out in remarkable detail a wide number of space related aspects (see figure 7.2); his later work is highly philosophical. A similar tendency can be noted in the works of another space pioneer, Hermann Oberth.

The mixture of science and philosophy in early Russia is often referred to as "Russian Cosmism", which dealt with the history and philosophy of the origin, evolution and future of the Universe and mankind [30]. Tsiolkovsky was heavily influenced by the Russian philosopher Nicolai Fedorov in this respect and in one of his later works of 1926, reflected this in a sixteen point "roadmap" for human space expansion as follows [30] p. 19.7:

1. Designing of rocket airplanes with wings
2. Progressively increasing the speed and altitude of these planes
3. Designing of a real rocket without wings
4. Ability to land on the surface of the sea
5. Achieving a velocity as high as 8 km/s, allowing the rocket to break through the atmosphere
6. Lengthening of rocket flight time in space
7. Use of plants to make an artificial atmosphere in spacecraft
8. Use of pressurised space suits for activity outside spacecraft

9. Making orbiting greenhouses for plants
10. Building of large orbiting habitats around Earth
11. Use of solar radiation to grow food, heat space quarters and for transport throughout the solar system
12. Colonialisation of the asteroid belt
13. Colonialisation of the entire solar system and beyond
14. Perfection of society and its individual members
15. Overcrowding of the Solar System and colonialisation of the Galaxy
16. The Sun begins to die and those people remaining in the Solar Systems go to other Suns

Now some 75 years later, we can only be astonished about the accuracy of the technical steps predicted (1 to 11). Space exploration may not have exactly followed this pattern but in broad terms the prediction has been extremely accurate (with some deviation on the overemphasis about growing plants in space). In view of this accuracy one could start wondering why not also the other steps will materialise once...

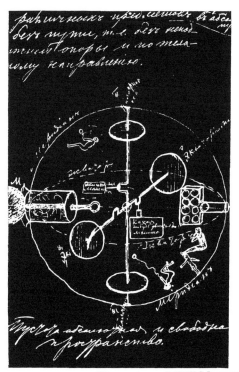

(By courtesy of ZPK Museum)

Fig. 7.4: Sketch of Tsiolkovsky (1883)

An interesting observation is made in this respect by Jesco von Puttkamer, the NASA scientist responsible for future projects. He points out that in the past science and philosophy were dealt with and taught by the same persons. Some 350 years ago a "schism" occurred under the influence of rational thinkers such as Descartes, Locke and Newton, splitting the material sciences and the spiritual world into two separate domains.

Such a split was necessary to allow scientists to develop their fields freely (cfr. the interactions during the times of Galileo) but also both fields have progressed considerably in knowledge leading to a form of "specialisation" in the 19th Century. From that point science moved forward in its material pursuits. Some form of joining may be necessary now, as expressed in [31] p.24:

Today this schism, this tremendous gap between the spiritual world and the material world of science, causes great concern to many people, particularly also to scientists. With everything in the physical sciences, specifically in subatomic physics as well as cosmology, becoming, by necessity, more statistical, there is a feeling that certain concepts from the spiritual side may have to be taken into consideration.
(…) I think it could be one of nature's principles that for each step further out into the physical world, we are also taking a step deeper into our inner world, whether we want it or not.

The latter statement was also worded by Socrates, as early as 450 BC:

Man must rise above the Earth – to the top of the atmosphere and beyond – for only then will he fully understand the world in which he lives.

It highlights a paradoxical paradigm at the first glance: mankind may have to go higher and higher to understand itself deeper and deeper.

In any case it is a philosophical dimension which obviously was recognised by the early space pioneers and was, afterwards, overruled by the materialistic space race, only to come back now when the schism between science and philosophy is closing again.

7.6 THE CROSS-CULTURAL COOPERATION IN SPACE PROJECTS

In this same category of philosophy we have mention the important dimension of international cooperation. More and more often, for example in the Space Station era, one notices the open discussions on the political dimension. The space activities have gradually evolved from a symbol of confrontation into one of cooperation.

CULTURAL ASPECTS IN INTERNATIONAL PROJECTS

Terpstra is probably the most recognised authority in the field of the cultural environment of international business. He defines culture as [32] p.36:

Culture is an integrated total pattern of learned behaviour shared by members of a society.

Or, culture is to a society what personality is to an individual. International contacts are influenced by a number of factors, which complicate this process. Terpstra distinguishes between [33] p.84:

- Material culture and standards
- Language
- Aesthetics
- Education
- Religion, beliefs and attitudes
- Social organization
- Political life

The major source of conflicts and communication problems during international contacts are due to the "self-reference" criterion and to the confusion when being confronted with a different culture which deviates considerably from this criterion. To illustrate some differences, we can quote some communication-related examples from [34]:

- The American OK sign (round finger) means zero in Europe, money in Japan and is even a vulgar gesture in Russia
- To say "no", people shake their head from side to side in Europe, jerk their head back in the Middle East, wave a hand in front of the face in the Orient and shake a finger from side to side in Ethiopia

- Whereas the colour black signifies death in many countries, white represents death in Japan and many Asian countries.

Terpstra illustrates the effects of cultural differences as follows [32] p.37

> *The fish is at home in, and comfortably unconscious of, its environment, which is water. The fish only becomes uncomfortable and aware of its environment when it is out of the water. In the same way, we are comfortably unaware of our cultural environment until we leave it and enter another culture. Then we, like the fish, experience what we call <u>culture shock</u>. Culture shock is the malaise we feel when we discover that there are other behaviours and values that may be accepted alternatives to ours. We are not sure of the "right" way to behave.*

In a confined environment such as on board of a space station where there are few alternatives, it is evident that these effects are amplified and the influence of these psychological effects on crew behaviour and performance has been studied in depth [35]. By respecting the other parties' cultural values problems can be compensated for and by having a sufficient knowledge of the foreign culture such tensions can be anticipated and the prime objectives achieved.

We can illustrate the factors mentioned by Terpstra with some examples from multicultural spaceflight:

Language

From of a questionnaire distributed to Russian and American astronauts, it was concluded [36]:

- There is a consensus that only one shared and common language will be used on board for technical conversations, with an increased importance according to the duration of the flight
- Dialects are considered less hindering
- Watching and listening activities increased in space, contrary to more complex communications such as reading and conversation
- Communications in space were hindered by environmental factors, such as ambient noise and facial swelling.

Material Culture

In general, Russian space systems and training are based more upon mechanical tools, whereas American and Western European rely more on electronic or other aids.

Examples of some different approaches are:

- American and Western European docking mechanisms are fully automated, Russian systems put more emphasis on manual override capacities
- Russian doctors strongly believe in mechanical preparation against space sickness (rotating chairs), whereas American doctors prefer to provide medicine against this syndrome.

Aesthetics

In French this point is easily solved, "les gouts et les couleurs ne se discuttent pas" (you cannot argue on taste and colours). In a confined space this is rather different; melancholically Russian music is not to everyone's taste, as the same is true for food. Even colours and decoration can have very different meanings as we can illustrate using the following examples:

- Blue is seen as a very masculine colour in most parts of the world, but so is red in France. However, Iranians will feel very uncomfortable in a blue dominated environment as it represents a "bad colour". On the other hand red is felt to be blasphemous in many African countries
- Flower decorations can be equally dangerous. White flowers are associated with death in Japan, whereas purple flowers symbolise death in Brazil and yellow flowers symbolise the same in Mexico and parts of Europe.

Education

In the early manned space activities, the crews consisted of pilot astronauts, mainly selected from a population of military test pilots. Therefore this group had a rather homogeneous education and background.

These crews are now more heterogeneous by the involvement of mission and payload specialists often with for example a scientific or medical background and education. The differences between the military, organised and disciplined education and the scientific, intellectually questioning, education is evident and a source of conflicts specifically in critical situations.

Religion, beliefs and attitudes

There are numerous aspects, which can be quoted here and imagined, simply by imagining the fact that Christians, Buddhist, Moslems and atheists one day will have to live and work in a closed environment in space. However, there are some less obvious aspects such as:
- Different perception on medical data; medical data in the Russian system are more public than for example in Western Europe.
- Background information and attestations during selection which are not consistent with Japanese cultural values.
- The need for privacy, influencing the crew quarters design.

Social organisation

The differences in this respect became obvious when the "Code of Conduct" on board of the ISS started to be discussed. Acceptance of leadership is a cultural factor which is based upon our social habits.

Political

To be ahead in space exploration has been viewed in the past as a primary means of achieving dominance in the power struggle of two competing systems. The symbolical value of great technological achievements served to demonstrate to the outside world power, vitality, and the prowess of a nation (cfr. The Chinese Wall, The Roman Aquaduct System, The Eiffel Tower).
Astronauts and cosmonauts were sent as political "ambassadors" in this belief. Evidently having to work now together is sometimes leads to conflicts.

CROSS-CULTURAL COUNTERMEASURES

Geert Hofstede, a psychologist interested in the various cultural differences of nationalities, started to specialise in "Xenopsychology" and based upon 116.000 questionnaires in 20 languages, "plotted" cultural parameters for 40 countries (later extended to 53).

The four parameters he finally concluded from these surveys can be described as follows: [37]

Individualism vs. Collectivism

Concerns the relationship between an individual and his/her fellow individuals. In certain societies ties are very loose and everybody is supposed to look after his own self-interest (e.g. Australia). The opposite of such an individualistic society is a society where people live "in groups" from birth (e.g. Pakistan).

Power distance

Refers to the fact that people are basically unequal, in physical and intellectual capacities. In some societies, where the power distance is high, such inequality is largely accepted and maintained (e.g. India), whereas in other societies a maximum effort is made to reach maximum equality (e.g. Israel).

Uncertainty Avoidance

Stems from the basic observation that the future is unknown and always will be. Some societies accept this uncertainty more easily than others do; one could qualify them as being more philosophical. One can also link this factor to conservatism in case of high uncertainty avoidance (e.g. Singapore).

Masculinity vs. Femininity

This relates to the social role of sexes in society. In masculine societies such role division is maximised (e.g. Japan), contrary to societies where amongst others women play a role in political life and other equal roles (e.g. Norway).

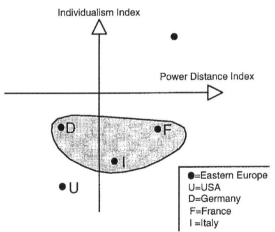

Fig. 7.5 Power Distance vs. Individualism

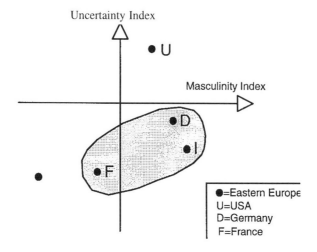

Fig. 7.6 Masculinity vs. Uncertainty Avoidance

Hofstede's cultural parameters are illustrated in figures 7.5 and 7.6. As an approach for the purpose of this topic, an "East European" point is plotted (●) compared with the related values of the three ESA nationalities France, Germany and Italy (F,D,I). The origin of the axes represents the average value resulting from Hofstede's results. Data for the U.S. are added as an additional reference point (U). If we attempt to convert Hofstede's results to the ISS we encounter the problem of heterogeneity on both the European and the Russian side. In the case of ESA, one is theoretically confronted with 15 different member states and thus with 15 different cultures. The same effect applies to CIS "nationals". To a Russian national a Kazakh has a culture as "exotic" as certain European ones.

Hofstede's cultural parameters are applicable to international negotiations but less to operational contacts that are based more on interpersonal relations and related communication problems. One should not fall into a quantitative discussion on such qualitative data but some relative conclusions may be drawn:

- Whereas Western European countries cluster relatively close together, the U.S. and CIS data are clearly distinct
- The Western European figures are located somewhere between both other ones

- The collectivistic dimension is certainly not a surprise. In Eastern Europe family relationships are much more expressive than in the more individualistic U.S.
- The social role division between the sexes is minimised in Eastern Europe. Women are allowed to take many different roles and men will play an active role in family activities. This is certainly the result of a political "equal treatment" doctrine.

Even if this approach is theoretical, reality has proven that a proper understanding and respect for each other's cultural values, long before an actual spaceflight have contributed considerably to the on-board psychological climate during the flight. The theory was put in practice for the Russian-European "EUROMIR" flights for the European crewmembers, including [38]:

- Integration in the Russian system during training
- Familiarity with Russian food
- Familiarity with Russian customs / private contacts
- Some knowledge of Russian culture and background.

These elements go much further than a good knowledge of the language which is only one of the many cultural parameters. However, as a side remark, also for supporting staff it is considered - for similar reasons - of utmost importance to master the Russian language at a basic level. As also in the technical dealings in Russia are governed by interpersonal relations and mutual trust, such knowledge considerably contributes to the mission success. Closer to practice is the book of Gesteland [39]. Here a distinction is made using the following parameters:

1. **Relationship-focused business cultures**:
   People prefer to do business with persons they are familiar with; indirect, polite, communication; lawyers play a consulting role.
2. **Deal-focused business cultures**:
   Clear business language preferred (no small-talk introduction); direct, frank communication; negotiations often led by lawyers.
3. **Formal, hierarchical business cultures**:
   Often "protocolarian rituals", formal interpersonal communication; status and titles are valued.
4. **Informal, egalitarian business cultures**:
   Informal behaviour is not seen as disrespectful, use of first names.

5. **Polychronic business cultures**:
   Schedules and deadlines are more flexible, meetings are frequently interrupted.
6. **Monochronic business cultures**:
   Punctuality is very important, schedules and deadlines are rigid; meetings are seldom interrupted.
7. **Reserved business cultures**:
   People speak softly; little physical contact and eye contact; few hand and arm gestures.
8. **Expressive business cultures**:
   People speak quite loud, physical touching and intense eye contact; vigorous hand and arm gestures.

If we make the link to space activities, we could for example distinguish the main ISS partners as per table 7.5

	U.S.	RUSSIA	JAPAN	GERMANY	FRANCE	ITALY
1. Relationship-focused		XX	XX			
2. Deal-focused	XX			XX	X	X
3. Formal, hierarchical		XX	XX	X	XX	X
4. Informal, egalitarian	XX					
5. Polychronic		XX				X
6. Monochronic	XX		XX	XX	X	
7. Reserved			XX	XX		
8. Expressive	X	X			XX	XX

(XX = strong, X = variable)

Table 7.5: Differences in business cultures of main ISS partners

From table 7.5 we can deduce:

- The ISS participants cover a wide scale of differences
- Also here, it is evident that Russian and American styles are almost opposite
- On average, Europeans are situated between both being largely a mixture.

CHAPTER 7 : SPACE AND PHILOSOPHY   251

PEOPLE WORKING TOGETHER IN INTERNATIONAL SPACE PROJECTS.

The above-described problems make the challenge even more interesting. One should not forget that not only the astronauts will have to learn to work together but equally, thousands of people preparing the project on the ground.

The joint Apollo-Soyuz project in 1975 was an unprecedented event in this context. Two nations who carefully had kept their technical skills secret from the other partner had to work together and, in view of safety and other interface aspects, were forced to exchange information.

The idea of the ASE, the Association of Space Explorers, originated from that period. ASE members are astronauts and cosmonauts who have made a spaceflight and who meet regularly, in this way exchanging ideas in an informal and non-political forum.

(Photo: NASA archive)

Fig. 7.7: The historical Soyuz-Apollo rendezvous in space in 1975

The aspect of "Working together in Space" is now accepted as a general motivation for every manned project. Specifically the International Space Station is considered as an interesting "social testbed" for this (chapter 8).

A lot of emphasis is put on the fact that people of different cultures working together will go through a number of psychological "phases" in the course of a project. Whereas this is described as a U-shaped curve in Xenopsychology, we can also relate this to the various project phases (adapted from [40] p.216):

	PHASE A/B			PHASE C/D
Psychological effect	Euphoria	Culture Shock	Cultural Assimilation	Stability
Effect on Teamspirit	Positive	Negative	Change from Negative to Positive	Positive

Table 7.6: Cross-cultural evolution over time in projects

In [17] this is expressed as follows:

Even so, the politically paramount consideration that it fosters peaceful cooperation between nations and serves to surmount antagonisms may well be worth the prize of a somewhat less than optimal efficiency of implementing projects – an experience to which, on a smaller scale, the European nations collaborating in ESA have provided ample evidence. There may even be the possibility for human exploration to space to become more than a vision for a single nation; it could become a means to effectively promote, perhaps even start to implement, the idea of peacefully united mankind.

One of the best illustrations of this philosophical dimension is what Yuri Gagarin wrote in 1962 on a picture after his spaceflight:

When I circled the Earth in my space capsule, I saw how beautiful our planet in reality is. People from the planet Earth, I hope together we will manage to preserve or augment this indescribable beauty, but in any case never to destroy it.

In the same sense, the 1977 Voyager space capsule carried the following message of hope "on behalf of the people who inhabit the planet Earth":

We human beings are still divided into nation states, but these are rapidly becoming a single global civilisation. (...) We hope someday, having solved the problems we face, to join a community of galactic civilizations.

Also in [17], the authors further plead to use openly this kind of political arguments to obtain public support for space programmes. They correctly point out that other arguments such as the production of materials in space can backfire considerably if the public realises that they were not realistic or cannot be achieved economically:

The resulting benefits are neither readily measurable in monetary terms nor do they belong to the scientific category. They may be characterised as intended and desirable side effects when viewed from a scientific or an economic perspective. Nevertheless, they have often been in the past, and still are today, determining factors in the decision processes for the initiation of new projects when economic or scientific justifications alone proved unconvincing. (...) These days, new political objectives such as international cooperation and the conversion of military industries into peaceful ventures are rapidly gaining importance as geopolitical conditions are changing radically and new necessities are resulting.

One day there will be a project to put a man on Mars and it cannot be sold as a mission that will bring back tonnes of precious materials (even if this "marketing" helped Columbus in discovering America). One should be open enough to underline the pioneering aspects of such a mission to the people; enthusiasm for this is large enough to find sufficient support without trying to sell aspects that could "backfire" sooner or later.

7.7 CONCLUSION

There can be little doubts about the philosophical dimension associated to space activities of an exploratory nature. This is the case for scientific projects where we are trying to find answers to the "origin of life" and the distant past of our planetary system in the expanding universe, but certainly for the purpose of manned space endeavours.

Exploration of other planets and celestial bodies is a logical continuation of mankind's drive to extend its frontiers. This is an inherent part of mankind's culture or, in a broader context, part of a natural anthropological process.

All space pioneers had this idea in mind, from Tsiolkovsky and Oberth to Von Braun. Rockets were transportation means to explore the Universe and space stations "technical" intermediate steps needed for logistics reasons.

This "frontier" drive appeals to a broad public and is the basic justification for a number of exploratory space activities, especially now that the philosophical and even theological community have accepted the rationale for space exploration also.

The vulnerability of the Earth to total destruction from the outside by the impact of an asteroid should be a sufficient argument already to continue this drive; surprisingly though, this is not the case.

A second major aspect in this context is the multinational span of big space projects which opens a new dimension in which people of different cultures and even opinions work together. Therefore the International Space Station can be seen as an important social testbed for human cooperation which may already constitute an important enough argument on its own.

I am convinced that the importance of these arguments justifies treating Philosophy, in this particular case, as a part of the "Space Marketing Mix" on its own, and not as a subset of the other more traditional tools.

REFERENCES CHAPTER 7

1. VON BRAUN, W., *Space Frontier* (Holt, New York, 1967)
2. AUGUSTINE, N., *Augustine's Laws* (Viking, New York, 1986) pp. 217-219.
3. PETERS, T. and WATERMAN, R., *In Search of Excellence* (Warner Books, New York, 1982), p. 42.
4. KUHN, T., *The Structure of Scientific Revolutions* (Univ. of Chicago Press, Chicago, 1970).
5. KOELLE, H., Chancen und Herausforderungen der Raumfahrttechnik im 21. Jahrhundert. In: SAHM P. and THIELE G., *Der Mensch im Kosmos*. (OPA, Amsterdam, 1998), p.94.
6. ESA, *Space Science and the Long-Term Future of Space in Europe* (ESA, Noordwijk, 1998) p.185.
7. KANTER, R., *The Change Masters*, (Thomson, London, 13th reprint, 1997).
8. MICKLETHWAIT, J. and WOOLRIDGE, A., *The Witch Doctors*, (Mandarin, London, 1997).
9. PASCALE, R. and ATHOS, A., *The Art of Japanese Management*, (Penguin. Harmondsworth, 1982).
10. OHMAE, K., *The Mind of the Strategist*, (Penguin, London, 1982).
11. POTTER, B., *The Way of the Ronin*, (Ronin publ., Berkeley, 1988).
12. MUSASHI, M., *The Book of Five Rings*, (Bantam, New York, 1982).
13. STEINER, C., A Philosophy for Innovation, *Journal of Product Innovation Management*, Vol. 12(5) (1995) pp. 431-440.
14. WITTMAN, J., Administrator Goldin Says NASA Needs Young Recruits. *Space News*, (March 27, 2000), p.10.
15. SQUEO, A., Attracting Top Engineering Talent Requires Some Creative Thinking, *The Wall Street Journal Europe*, (25 April 2000) p.30.
16. KANTER, R., From Spare Change to Real Change. *Harvard Business Review*, (May-June, 1999) pp. 122-132.
17. SAPHIR, Dokumentation des Workshops am 3. Juli 1991 by der DLR in Koeln-Porz, (DLR, Cologne, September 1991).
18. FROMM, J. and HOEVELMANN, G., Technology Assessment of Human Spaceflight: Combining Philosophical and Technical Issues. *Paper presented at World Space Congress 1992,* (IAF, Washington, September 1992).
19. KERNER, M., Der Mensch in der Geschichte, in SAHM, P. and THIELE, G., *Der Mensch in Kosmos,* (OPA, Amsterdam, 1998), p. 9.
20. ZABUSKY, S., *Launching Europe. An Ethnography of European Cooperation in Space.* (Princeton University Press, Princeton, New Jersey, 1995).
21. SAHM, P. and THIELE, G., *Der Mensch in Kosmos*, (OPA, Amsterdam, 1998).
22. TIPPLER, F., *The Physics in Immortality*, (Doubleday, New York, 1994)
23. TEILHARD DE CHARDIN, P. *Le Phénomène Humain*, (Ed. Du Seuil, Paris, 1955).
24. LEIBNIZ, G., *Monadologie* (1714), par. 53 (Hamburg, 1956).
25. UN, *Treaty on Principles Governing the Activities of States in the Exploration and Use of Outer Space, including the Moon and other Celestial Bodies*, (UN, London/Moscow/Washington, 10 October 1967).
26. SOLEM, J., Deflection and Disruption of Asteroids on Collision Course with Earth. *Journal of the British Interplanetary Society* Vol. 53 (2000) pp. 180-196.
27. JOHNSON-FREESE, J. and KNOX, J., Preventing Armageddon : Hype or Reality. *Journal of the British Interplanetary Society* Vol. 53 (2000) pp. 173-179.
28. SAGAN, C., Opening Keynote Address of 'What is the Value of Space Exploration?' *NASA Symposium*, (NASA, Washington D.C., July 1994) p. 2.

29. LÜST, R., *Europe and Space*, ESA-BR-35, (ESA, Noordwijk, 1987) p.23.
30. FINNEY, B. and LYTKIN, V., Origins of the Space Age: Tsiolkovsky and Russian Cosmism; in: HOUSTON, A. and RYCROFT, M., *Keys to Space*, (McGraw Hill, Boston, 1999).
31. VON PUTTKAMER, J., Foreword in HARRIS, P., *Living and Working in Space*, (Ellis Horwood, New York, 1992), p.24.
32. TERPSTRA, V., *International Dimensions of Marketing*. (Wadsworth, 3rd Ed., Belmont, 1993).
33. TERPSTRA, V., *International Marketing*, (Holt, New York, 1972).
34. RICKS, D., *Big Business Blunders: Mistakes in International Marketing*, (Irwin, Homewood, 1983).
35. CASSIDY-CURTIS, T., Psychological Challenges of Manned Spaceflight. *Journal of the British Interplanetary Society*, Vol. 46 (1993).
36. KELLY, A. and KANAS, N., Crewmember Communications in Space: A Survey of Astronauts and Cosmonauts. *Journal of Aviation, Space and Environmental Medicine*. (August 1992), pp. 721-726.
37. HOFSTEDE, G., *Culture's Consequences*, (Sage Publ., London, 1980).
38. PEETERS, W. and SCIACOVELLI, S., Communication related aspects in Multinational Missions, *Journal of the British Interplanetary Society*, Vol. 49 (1996), pp. 113-120.
39. GESTELAND, R., *Cross-cultural Business Behaviour*, (Copenhagen Business School Press, 1999).
40. BOUTELLIER, R., GASSMANN, O. & VON ZEDTWITZ, M., *Managing Global Innovation*, (Springer-Verlag, Berlin, 1999).

Chapter 8

CASE STUDY: INTERNATIONAL SPACE STATION COMMERCIALISATION

8.1 INTRODUCTION

The International Space Station (ISS) is undoubtedly the most prestigious and perhaps also the most criticised space project of our era. A permanently manned station has been projected by all the main space pioneers as a stepping stone for further space exploration. An overview of the different plans is given in table 8.1. From this overview we can also deduce that the political motivation for ISS has evolved considerably over the last decade. At present, the project has gained a momentum which cannot be stopped abruptly, but the fear for cost increases is forcing us to think about the financing of the exploitation phase.

Therefore, ISS will undoubtedly become a precursor of a global PPP concept. Private investors were not involved in the design phase and the station is being built on the basis of Agency requirements and public funding.

However, the operational cost is now under constant scrutiny and therefore private investors are being approached to participate in the utilisation of ISS. In view of this, all previously mentioned problems, specifically with respect to Intellectual Property and access, are now evident and have to be solved in the next few years.

This touches all the marketing elements previously discussed and makes ISS an excellent and realistic case study.

1869	• E. Hale described in the "Atlantic Monthly" a station called the "Brick Moon", with a diameter of 60 m and a permanent crew of 37, to help navigation of ships
1911	• Tsiolkovsky lays foundations for space travel and space stations
1923	• In his book "The Rocket to Interplanetary Space" H. Oberth realised that, for a Mars voyage, a "refuelling point" would be needed in outer space, introducing the term "Weltraumstation"
1936	• In another pioneering work "Rockets through Space" P. E. Cleator describes a station in detail, locating it at a height of 600 miles. Even costs were calculated and it was reckoned that the project could be realised for less than 10 million $
1952	• A rotating wheel concept, with a diameter of 85 m, triple-decked and partially inflatable was worked out by W. von Braun
1971	• Russia launched the first of a series of Salyuts. A total of seven Salyuts were put in orbit (13 m long and 3.6-4.2 m diameter)
1973	• Skylab (36.1 x 28 m) was launched as an interim phase towards a permanently manned station and orbited the Earth for six years (with, however, only 171 days of operations)
25.1.1984	• President Reagan directs, in his State to the Union address, NASA to develop a permanently occupied space station within a decade and invites other countries to participate in the project
20.2.1986	• The core module of the Russian MIR station was launched. The station was gradually upgraded up to 1996 (Priroda module)
29.9.1988	• Multilateral intergovernmental agreements are signed between the U.S., ESA, Japan and Canada (the "International Partners") for the building of Space Station Freedom
25.1.1994	• President Clinton takes an important step, he notes in his State of the Union that "Instead of building weapons in space, Russian scientists will help us to build the international space station". The station was now called Alpha
20.11.1998	• The first element of the now called ISS, Zarya (stands for "dawn" in Russian) is launched with a Proton rocket. Two weeks later Node 1 (Unity) is brought up by the Shuttle and attached.
27.7.2000	• The third ISS element, Zvezda ("star") is successfully docked.

Table 8.1: Historical overview of Space Stations

Detailed descriptions of this evolution can be found in [1], [2] and especially [3].

8.2 SHORT DESCRIPTION OF THE PROGRAMME

The Space Station design is an iterative process and will undergo still a number of changes in the coming years. As a baseline and a description of the main design parameters the book of Messerschmid [3] has been taken. This book represents an excellent overview of Space Stations in general.

The basis of the ISS is a large structure, called "Truss", of approximately 100 m length which carries 4 pairs of large photovoltaic solar power generators, several arrays of thermal radiators to remove the heat generated on board the Station into space and a combination of pressurised modules.

The Station will be outfitted with six pressurised modules, namely:

- The US Laboratory module, called "US Lab"
- The Japanese module, called "JEM" (Japanese Experiment Module)
- The European Laboratory Module "Columbus"
- Three Russian research modules.

"Columbus" is a cylindrical module (see figure 8.1) and is Europe's main contribution to the ISS.

There are a number of external payload sites (4 U.S., 1 Japanese, TBD Russian) and external activities will be supported by a "Space Station Remote Manipulator System", which is a Canadian contribution. A second robotics arm is mobile c.q. movable, called the "ERA" (European Robotics Arm) and represents the second European important contribution.

The main characteristics of the present International Space Station design are summarised in table 8.2

Based upon [3], a comparison is presented in table 8.3 between space stations, where it becomes quickly evident that ISS will combine the mechanical features of MIR with the data and power management aspects developed in the Western world.

(picture ESA/D. Ducros)

Fig. 8.1: Artist impression of ISS with ESA Columbus module

DIMENSIONS	108.4 m x 74.1 m
MASS	415.000 kg
ELECTRICAL POWER	110 kW of which 47-50 kW for research work
PRESSURIZED VOLUME	1140 m^3
CREW	3 (1998-2002), 7 (from 2002 onwards)
ORBIT	51.6 deg , circular orbit, varying altitudes (335-460 km)
MICROGRAVITY LEVEL	10^{-6} up to 0.1 Hz; 10^{-5} up to 100 Hz.
UNDIST. MICROGR.	30 days

Table 8.2: ISS Main Characteristics

	STS + SPACELAB	MIR	ISS
MASS	13.700 kg	140.000 kg	415.000 kg
DIMENSIONS	6.9 x 4.1 m	33 x 41 m	108 x 74 m
PRESSURIZED VOLUME	166 m^3	410 m^3	1140 m^3
CREW	Up to 7	3 (perm.)	Up to 7 (perm.)
SOLAR ARRAY AREA	0	430 m^2	3000 m^2
USER POWER	3.5-7.7 kW	4.5 kW	47-50 kW
DATA TRANSMISSION	45 Mbps (down)	7 Mbps (down)	50 Mbps (down)

Table 8.3: Comparison of Space Stations

As already mentioned in the general description, the main ESA contribution is the Columbus module, a cylindrical module with an overall length of 6.7 meters and an external diameter of 4.5 meters. The module will be permanently attached to the International Space Station from which it will receive power and other resources. The projected lifetime is 10 years and it can be easily reconfigured by the exchange of standardised instrument carriers, called International Standard Payload racks (ISPR). The main characteristics of the Columbus module are given in table 8.4.

VOLUME	6.7 m long, 4.5 m diameter
LAUNCH MASS	12700 kg
PAYLOAD PER RACK	400 kg
POWER	13.5 kW for Payload
VACUUM/VENTING	> 1.3 x 10^{-6} bar

Table 8.4: Columbus characteristics

Further contributions from Europe are as follows:

- Delivery of an external Robotics arm, called ERA
- Provision of Ariane5 services, coupled with a resource carrier called ATV (Automated Transfer Vehicle) which performs Station "reboost"

functions in addition to the regular resupply capacity. In view of the importance in the exploitation programme, the main characteristics are given in table 8.5.

VOLUME	10.1 m long, 4.5 m diameter
LAUNCH MASS	Up to 20500 kg
CARGO	Up to 7500 kg
SUPPLY CARGO	Up to 5500 kg dry cargo, 940 kg water/gas, 860 kg fuel
REBOOST CARGO	Up to 4000 kg propellant

Table 8.5: ATV characteristics

- In addition to this, ESA will deliver a number of so-called "Early Contribution Items", namely

   ** A Glovebox
   ** A Freezer ("MELFI")
   ** Hexapod
   ** A Mission Data Base

The purpose of this latter group of contributions is to place them in modules, which will be operational before the European module, this way having access to earlier utilisation opportunities during the construction phase of the Station.

- A Crew Return Vehicle (CRV) is developed as an inter-agency programme, also called X-38, with European participation from ESA and DLR (see earlier figure 1.3).

- Other hardware is delivered by Europe for the ISS in the context of "barter agreements" (instead of cash payments), namely

   ** Nodes 2 and 3
   ** Cupola

8.3 PRESENT COMMERCIALISATION EVOLUTION IN NASA AND CSA

NASA developed early plans for this in 1998 with the objectives [4]:

<u>Long term</u>: To establish the foundation for a marketplace and stimulate a national economy for space products and services in low-Earth orbit, where both demand and supply are dominated by the private sector.

<u>Short term</u>: To begin the transition to private investment and offset a share of the public cost for operating the space shuttle fleet and space station through commercial enterprise in open markets.

Essentially, it has been proposed to re-establish the Customer – Supplier Relationship by forming a Non-Government Organisation to act in the long run as the prime customer. In order to establish how to implement this approach, a study [5] was performed which concluded that the five most viable options for managing the US elements of the ISS were, in order of independence from NASA:

- NASA Institutes
- Consortia
- Government Corporations
- Government Sponsored Enterprises
- Cooperative Associations.

A number of enabling legislative steps, applicable for most cases are suggested (such as liability waivers, excemption from standard procurement regulations etc.). Formally, the road was prepared by the Commercial Space Act of 1998 (Public Law 105 – 303), with a specific section 101 on the "Commercialisation of the Space Station", stating that:

> *The Congress declares that a priority goal of constructing the International Space Station is the economic development of Earth orbital space.*
> *The Congress further declares that free and competitive markets create the most efficient conditions for promoting economic development, and should therefore govern the economic development of Earth orbital space.*

> *The Congress further declares that the use of free market principles in operating, servicing, allocating the use of, and adding capabilities to the Space Station, and the resulting fullest possible engagement of commercial providers and participation of commercial users, will reduce Space Station operational costs for all partners and the Federal Government's share of the United States burden to fund operations.*

On the basis of this NASA created in February 1999 a special office for Commercialisation. The newly appointed "Assistant to the Administrator for Commercialisation" is in charge of seeking opportunities to increase Commercialisation of NASA infrastructure, operations and technology. Evidently, this is not exactly a "novelty", but the underlying aspect is an attempt by NASA to "counter the anti-commercial culture" [6].

This does not mean that NASA has no experience in commercialisation; earlier NASA commercialisation efforts inter alia include [7]:

- 1962: NASA spun off control of communication satellites to Comsat
- 1996 : VentureStar (X-33), joint development with Lockheed
- 1996: United Space Alliance (a Lockheed/Boeing Joint Venture) was awarded the SFOC (Space Flight Operations Contract), a contract for Shuttle operations, with a value of 8.5 billion $ (six years commitment), presently employing some 10,000 people
- 1999 : Litton/PRC takes over NASA's sounding rocket operations (up to some 600 million $ over 10 years)
- 1999: Lockheed wins a 3.4 billion $, 10 year contract to manage all NASA's communications, data collection and telemetry operations. This CSOC (Consolidated Space Operations Contract) merges 15 separate contracts under one single management structure, with a team of 40 companies (expected savings: 1.4 billion $).

With the backing of the a.m. legal support, the next target is clearly a Commercialisation of the Space Station operations. Goldin [8] has put as a target the initial commercialisation 30 percent of the U.S. share of the Space Station internal resources. As the commercial development programme would mature, he is even prepared to increase this figure considerably:

> *The private sector will decide how to utilize this [30%] set-aside. This means it is now time for private business, big and small, to plot their strategies to utilize this world-class asset.*

Nothing would please me more than if commercial demand for Station accommodations reached 40, 50 or even 80%. Sound surprising? Let me assure you NASA would be happy to be a minority tenant in this facility.

Some people go even much further. Tumlinson [9] suggests to NASA:

- Turn over the Shuttle fleet to private operators, allowing them to use "spare capacity" (such as 5-10 million $ paying passengers)
- Privatise the Space Station by handing it over to a private multinational consortium
- Take the savings of these privatisations and apply them to opening of Mars and the rest of the solar system to human exploration.

We have to mention here that a study carried out by U.S. industry revealed a rather weak interest from their side, the conclusions on critical impediments are summarised as follows [10]:

- Uncertainty and magnitude of total price (i.e. transportation, integration and operations) makes return on investment (ROI) impossible to calculate for potential commercial users
- The lack of awareness in the non-aerospace community regarding the defining characteristics of the ISS (capabilities, range of potential uses and values) among potential participants inhibits serious market demand for those characteristics
- The restrictions in existing law, regulations, and policy, which limit commercial payloads on Shuttle limits prospective interest in ISS commercial use
- The burden of complex rules and procedures associated with accessing the ISS via piloted or non-piloted launch vehicles have deterred potential commercial users because of unacceptable cost and schedule penalties
- Lack of guaranteed Space Shuttle access on a regular basis to commercial users due to lack of availability of the Space Shuttle on a predictable schedule deters potential industry users that require multiple flights of experiments within a defined period.

Indeed, early analysis showed that in the initial phase of the ISS activities only limited industrial activity will be attracted. This is due also to the unsolved questions on pricing and Intellectual Property matters and the lack of guarantees in the transitional phase. On the other hand, a considerable interest was noted from the "entertainment" and media side, which also has led to the Enterprise project of Spacehab. It is assumed that this interest will be considerable in the first years of the station, but then will gradually fade out and resources will be redeployed for industrial applications.

Seeking active commercial interest, NASA found out at the end of 1999 that there were only a few potential pathfinder areas commercially proposed and therefore decided, in an unprecedented move, to enter the area of space imagery, including education, entertainment and even advertising. The "Dreamtime" project was launched in June 2000 (see: www.dreamtime.com) and involves roughly half of the NASA resources made available for commercial applications (i.e. 50% of the earmarked 30% commercial resources).

Industry is implementing the project and the private contribution is estimated to be in the order of 100 million $ (a seven years agreement with a five years option is foreseen).

The Agreement is described as a collaborative effort whereby each partner brings in resources for the joint undertaking. The commercial aspect is covered by the fact that NASA acknowledges that the material may be used to maximise commercial return, whereby also advertising and sponsoring are in principle allowed for.

The Canadian Space Agency (CSA) has opted for another approach. They intend to create a Canadian ISS Access Company (CIAC). The CIAC will be a fully commercial entity mandated to use the Canadian resources and licensed to sell the access to potential users [11].

While CSA brings in the access rights, the CIAC takes on the financial risk and capital investments. In return, the CIAC gets a relatively unrestricted right to make profitable business (so also here sponsoring and advertising are not excluded), within the boundary conditions of the binding Governmental Agreements. It is assumed in the pricing policy that in a later phase, when business is running in a profitable way, the CIAC will recompensate the Canadian Space Agency via royalties (as a percentage of the sales) or other fees. This income will then serve the CSA budget again.

8.4 ISS COMMERCIALISATION IN EUROPE

Similarly, from a European industry point of view, it has been pointed out [12] that a number of concerns will have to be settled first, namely

- The overall duration between the industry request and the mission results
- The overall cost including the development phases
- The legal aspects and property rights.

It has been mentioned already that data distribution will pose new challenges. With the long-duration involvement and interactivity in experiments, the experimenter, and even more industry, wants access during the flight operations to the data produced by his payloads. Moreover, he wants these data in his premises, also for integrity and data protection reasons. The cost aspect is obvious; before entering the system, industry will make a business plan. For this they need to know in advance

- Do they have to share in the launch cost?
- Do they have to pay for resources on board and in mission operations?
- What standards will be applicable and hence what qualification costs?

The only valid International Agreement so far is the previously described "Outer Space Treaty" [13], which was approved in 1967. Basically the problem is that commercialisation at that point in time was beyond the scope of imagination and therefore the treaty reflects a more normative, even ethical position. Although the legal relationships between ESA and NASA are clearly defined [14], the problem remains in a lack of appropriate laws.

INDUSTRIALISATION OF THE EXPLOITATION PROGRAMME

Until now we have considered the commercialisation aspects, but we cannot ignore different schemes of transferring from public to private interest. Formally, the following definitions are used:

> *Commercialisation:*
> *involves a private sector profit-seeking entity, using its own or borrowed and/or invested funds to carry out activities intended sooner or later to result in products or services that can be sold at a profit through a market, either to government or non-government customers, or to a mixture of the two.*

Privatisation:
involves a private sector profit-seeking entity, carrying out functions previously the responsibility of government.

Industry is not willing or able to finance basic infrastructure related to ISS. Moreover, the legal complexity with Intergovernmental Agreements will not allow such a scenario easily. On the other hand, for well-defined tasks or services which do not require such basic investment, industry is willing to take over the responsibility, be it under the umbrella of governmental guidelines. Such a concept, here called industrialisation, is appearing in the ISS environment. Traditionally, industry is more interested in continuous operations rather then one-shot developments. The estimated costs per phase for ISS are represented in table 8.6, including the first 10 years of operations [15], p.938.

PHASE	COST
A/B, Feasibility and design	4%
C/D, Development and Construction	45%
E, Exploitation	51%

Table 8.6: Cost distribution per phase for ISS

Although the Exploitation phase is drafted up for up to 10 years, the designed lifetime of for example the Columbus module is 15 years. Experience with other stations, such as MIR, have proven that with appropriate maintenance, lifetimes in the order of 1.5–2 times the designed lifetime are more usual. From this point of view, industry is showing an increasing interest in the exploitation programme. In an article [16] the modus operandi is clearly spelled out:

> *As soon as the Space Station is completed, NASA [ESA] should hand over its management to the private sector – to companies and people who understand contracts, time constraints and the bottom line. The Space Station partners should form an international authority like those that operate seaports and airports. This body would lay out the guidelines for commerce between the Space Station and Earth and amongst the Space Station tenants, enforcing laws on everything from intellectual property rights to rules governing disputes among tenants. It would then contract out the job of managing the Space Station's property.*

If we look into this exploitation phase, the utilisation cycle of the European part of the Space Station can be represented as per figure 8.2.

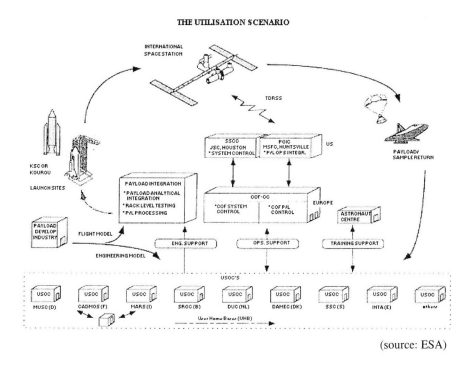

Fig. 8.2: The European utilisation scenario

We can distinguish a number of functions

- Operations Management, such as Programme planning, Mission management and coordination of European users
- System Operations, covering flight operations in the European Control Centre, Ground Communications, Engineering support, Logistics Support, ATV manufacturing
- Payload Operations, both in the integration as flight phase
- Astronauts and Crew Operations, covering astronaut's training and simulators, crew and medical support.

In this context

- A number of 8-10 logistic ATV missions (including the - for industry interesting - recurrent production of flight models) are at present foreseen till the year 2013
- An average yearly budget of 250 MEURO is foreseen in the ESA budget for steady state operations
- The role of the Agencies for the use of the European Ground segment has been settled.

In the beginning of 1999 a European industrial initiative was started by DASA to investigate the industrialisation of the European ISS operations. Alenia, Matra Marconi and DASA signed a Memorandum of Understanding and other companies showed an interest as well. The creation of Astrium reinforces the legal framework of this initiative. The contractual set-up foreseen presently will be based upon an O&U (Operations and Utilisation) Consortium, with a joint programme team composed from staff of the consortium members.

It is important to mention here the role of a Commercial Agent as part of the O&U Consortium, who will make the link between industrial customers and this ISS O&U or between ESA and the industrial customers (different options are still under consideration)

A second component decided by ESA involves various measures to encourage the commercial sector to buy up to 30% of the European Partner's share of utilisation capacity on the station. Recognising the complexity of such endeavour and taking into account the Inter Governmental and commercial agreements, the strong necessity for the Partners to mutually discuss the case for regulating the commercial practices adopted for the ISS environment strongly arises [18].

8.5 THE ISS PRODUCT

In essence, the unique commercial "product" offered by the International Space Station is a Microgravity testbed or laboratory for scientists as well as for industry which will :

- Improve industrial products and processes on Earth using the results of experiments done on ISS
- Introduce new commercial products and instruments to the market by licensing new technologies and methods used on ISS

- Use the Station as a platform to introduce and sell new services, for example in communications and imaging.

Certainly there are other means for providing Microgravity conditions but the primary advantages for the ISS users result from the following features:

- The primary advantage over other space systems is continuous availability over more than 10 years of exploitation, allowing proper planning
- This also permits for long-duration research activities, which will add a new dimension to research in space
- A number of resources available on board are considerable higher than those offered by other permanent stations or facilities before (for example power and data rates)
- There will be a permanent crew on board to cope, in general, with any unforeseen or unforeseeable behaviour of experiments within a short delay
- For ground-based scientific research, there will be a unique opportunity to modify or adapt research based upon results achieved in earlier steps.

From [3] p.310, we can deduce the main differences as per table 8.7.

SYSTEM	DURATION	P/L MASS (IN KG)	USER POWER (IN KWH)	MICROG. LEVEL
Droptower	4.5 s	125	0.6	10^{-5}
Parabolic Flight	20-30 s	50	2	10^{-2}
Sounding Rocket	6-15 min	30-80	1	10^{-4}
Spacelab	10-14 days	290	2.5	10^{-4}
ISS-Columbus	Years	400	3-6	$10^{-3} - 10^{-6}$

Table 8.7: Comparison of experimental conditions

From these considerations, the MIR station is comparable with ISS. However, There are a number of logistic constraints (such as power resources, data transmission and download capacity) which considerably put the balance in favour of ISS.

POTENTIAL APPLICATIONS

The list of potential users can only be forecast and will certainly grow over the years. Different classifications are made by different organisations, as shown in table 8.8.

ESA GROUPING [18]	NASA GROUPING [10]	EUROSPACE GROUPING [19]
• Biotechnology and biomedicine • Fluid thermodynamics • Combustion • Solidification Processes • Crystal Growth • Protein Crystallisation • Fundamental physics	• Biotechnology • Space technology testbed • Materials and processes • Entertainment • Education • Advertising	• Commercial Technology demonstration • Services demonstration • Agency technology demonstration • ISS enhancement demonstration • Space Station users services

Table 8.8: Classification of potential ISS applications

Although strictu sensu not comparable, the main purpose of putting these classifications in a "comparative" way is to illustrate the different viewpoints of public and private organisations.

INDUSTRIAL AND COMMERCIAL APPLICATIONS

The interested partners who are mainly industrial companies, are now being encouraged to come forward and propose ideas for using the Station as a laboratory. In order to do so, a number of "User Guides" are being developed by ESA mainly describing the characteristics and other boundary conditions.

Specifically from one of these [18] we can deduce following table 8.9, covering generic topics presently pursued in Europe.

DISCIPLINE	INDUSTRY RESEARCH	APPLICATION EXAMPLES
Casting of High-performance Alloys: • Advanced process control • Particle Reinforced Composites	• Cast products • Metal/matrix composites	• Turbine blades, car bearings • Reduced weight for cars and airframes
Crystal Growth: • Electronic/photonic • Biological macromolecules	• GaAs, ZnSe, CdTe crystals • Drug/active molecule interaction	• High sensitivity X-ray detectors • Faster drug design (HIV, tumors)
Particle Technologies	• Production of nanoscale particles	• Advanced thin layers (lenses, microchips)
Energy production and management: • Heat and mass transfer • Combustion	• Better understanding • Accurate models	• Oil recovery • Lower consumption and pollution of engines
Biotechnology and Medicine: • Biology and physiology • Tissue engineering • Health care	• Gene and cell behaviour • Tissue development, bioreactors • Preventive/therapeutic countermeasures	• Drugs modulating cell activities, • Artificial organs and biomaterials • Osteoporosis, Wound healing

Table 8.9: Generic table of ISS industrial applications

To illustrate these generic topics, a number of US companies are presently engaged in some of these themes, such as:

- 3M, on the development of thinner films
- NRC, molecular-beam epitaxy to produce pure semiconductor layers
- Paragon, better permeable contact lenses
- Coca-Cola, fizzy drinks,

just to quote a few.

ISS due to its high inclination is able to observe 85% of the Earth's surface on which 95% of the Earth's population is living. Therefore, it is equally important to consider the Station as a technology demonstration laboratory, where associated applications can be tested and validated with the help of manned assistance on board. A number of examples, deduced from [18], can be described as follows:

- **FOCUS:** an intelligent Fire Detection Infrared System on board of ISS, which will be able to determine the location and extent of fires, their nature, temperature range and pollutants emitted. Similar information can be provided for volcanic eruptions and will bring to respectively fire brigades and civil protection services early and analytical information.

- **SAGE:** the Stratospheric Aerosol and Gas Experiment, which will contribute to better understanding of the role of certain compounds in climatic processes, biochemical cycles and atmospheric chemistry.

- **ALADIN:** Atmospheric Laser Doppler Instrument, a Lidar application, measuring the altitude profile of winds on a global scale. This will improve knowledge on climate and, eventually, weather forecasting.

- **GTS:** Global Transmission Services; highly accurate time and data signals are transmitted, coded according to the specific requirements of the user on the ground. This can serve accurate time receipt for mobile users (on ships) and even car theft protection devices (coded electronic signals).

- **ACES:** Atomic Clock Assembly in Space; improving the accuracy of time measurement with a factor 100. This will lead to better synchronisation of intercontinental clocks but also to improve accuracy of such time-linked services such as GPS.

All these examples are based upon present industrial interest, but we can safely assume that many areas, presently not even considered, will appear during the operational lifetime of the Station. Therefore, dividing the areas in three categories provides a more systematic outlook [3] p.297:

1. Areas which are based upon earlier microgravity experiments and have aroused industrial interest, such as

   - Components for automobile and aerospace industry
   - Self-lubricating bearings
   - Electronic composites crystal growth (X-rays)
   - Protein crystals.

2. Areas subject to present research on Earth and in space, for which industry has outlined objectives. In this category fall:

- Osteoporosis treatment, wound healing
- Biotechnical processes such as electrophoresis, in order to improve efficiency on earth
- New methods for materials processing such as aerosols, multiphase fluids
- Investigation of combustion processes, in order to increase efficiency and, at the same time reduce pollution.

3. Future technologies that are assumed to be decisive for the competitiveness of process industries. Most important progress can be expected in:

- Information technology (higher storage capacity and quicker data transfer)
- Biotechnology (manipulate living systems at molecular level)
- New materials and manufacturing methods (micro-structures in the nanometer range).

The advantage of this qualification is that it underlines the evolutive and interactive process that will take place the next decade, with increasingly new applications building further on previous results. A key problem for all the above mentioned cases remains the transport cost of, say 20000 $ per kg. Only very few products will be so valuable in small enough quantities to reach a breakeven point. One of the rare examples is described [20] for alpha-interferon, a protein used to treat hepatitis and other disorders. Four production flasks could, according to the developing company, meet the 750 million $ yearly demands for the drug.

It is of interest to summarise those experiments in the literature which are linked to a potential commercial return. It is evident that this list is neither exhaustive nor complete because nearly each application should in fact be quantifiable (even with a large uncertainty range). Therefore table 8.10 is only exemplary but at the same time, it is hoped that it will encourage other disciplines to look into financial figures of this type.

RESEARCH AREA	PROPOSAL	COMMERCIAL POTENTIAL	SOURCE
Combustion	Efficiency of burners	2% efficiency = 8 billion $ saving in US	(21)
Osteoporosis	Countermeasures	10% countermeasure = 1 billion EURO in Europe	(18)
Proteins	Production of alpha interferon	750 million $ yearly	(20)
Technology demonstration	Testbed for commercial satellites	30-120 million $ yearly	(10)
Semiconductors	Material savings for IC-production	5% saving = 100 million $ yearly	(10)
Overflow sensor	Air conditioners condensation spilling	100 million $ yearly damage U.S. savings	(22)
Telemedicine	Diagnostic kit	50% less emergency landings = 500 million $ savings world-wide	Ch. 3

Table 8.10 Commercial potential of some ISS applications

As mentioned earlier, medical applications have a sensitivity aspect attached to them. However, we cannot ignore the interests of the pharmaceutical industry when dealing with medications applicable to a considerable part of the population. As described in [3] p.276, the Bone Mass Loss over eight months in space is equivalent to a ten years average Bone Loss in Humans from age 50-60. Therefore ISS will become the most appropriate testbed for pharmaceutical examinations or other (mechanical) countermeasures. However, there is a second macroeconomic aspect associated with this example. In Europe we estimate over one million fractures as a result of osteoporosis per year, a figure that will increase with increased longevity of the population. It is evident that this is a considerable direct cost not only for the patients but also for society in general (as already mentioned, the current treatment cost in Europe is 27 MEURO and is expected to increase with 65% by 2025).

However, as mentioned earlier, industry often has a different viewpoint on commercial applications; in principle they need a quick return, at least to recuperate soon the investment made (cash-flow management). In this context, what is announced as the "first commercial deal for the Station" follows a completely different approach [23]. Spacehab (U.S.) and RSC Energia (Russia) backed by partners such as DASA (Germany) and Mitsubishi (Japan) created a joint venture in December 1999 to attach the Enterprise module to the ISS. Enterprise is a 7 tonnes pressurised space module, 10 m long and approximately 4 m in diameter. It will house a broadcast station and research laboratory in which company-sponsored Microgravity experiments (targeting biotech and material sciences) will be conducted. Early funds, in the order of 100 million $, will be mainly collected from media companies whereas RSC Energia will participate by building and launching the module.

(By courtesy of SPACEHAB Inc.)

Fig. 8.3: The commercial Enterprise Module

8.6 PRICE OF THE STATION

NASA officially [21] maintains a cost estimate for the development cost as:

Cost Item	Up to 1998	Cost to Go	Total Cost
Development	8.9	3.1	12.0
Operations	0.8	4.0	4.8
Research	1.1	2.2	3.3
Russian Programme Assurance	0.3	0.9	1.2
Crew Return Vehicle	0.0	0.8	0.8
Total	11.1	11 – 13	22 - 24

Table 8.11. NASA development cost for ISS (1998 billion $)

An additional 13 billion $ are estimated for the operational cost over a 10 years lifetime. If we add to this an estimated 10 billion $ development contribution from the International partners for ISS and a similar amount for the operational phase, then the total ISS cost can be estimated to be in the order of 55-60 billion $.

From a technical point of view, we have noted from table 8.7 that ISS will provide a good microgravity environment which is available for a longer period of experimentation than some alternative methods. This has resulted in some authors [24] expressing the cost of these various options as the cost of microgravity per kg-hour. Based upon the costs per kg of payload and dividing this by the exposure time, they come to the relative results as per table 8.12.:

SYSTEM	RELATIVE COST PER KG-HOUR
Sounding Rocket	6,635
Parabolic Flight	28
Space Shuttle	9.5
Space Station	1

Table 8.12 Relative cost for Microgravity Experiments

Pricing policy

A considerable concern has been expressed by potential European users on the pricing policy which will be adopted. Based upon various remarks expressing this concern, Eurospace [19] draws following conclusions:

- ESA must define the rules to be applied worldwide for subsidising the commercial experiments to be selected
- These rules would benefit from being identical for all International Partners, to avoid competition
- The level of subsidisation must be sufficient to motivate industry to come aboard the ISS
- It is important that the process of co-funding by ESA or the other partners be transparent.

Summarising this as follows:

If the principles are fair and reasonable, then the commercial utilisation of the ISS will be a success. If the rules are unclear or not transparent, or vary for different users of the ISS, then some difficulties will occur, since it could even be considered as an infringement of the general rules of fair competition of the World Trade Market.

NASA has followed a structured approach. Based upon Public Law 106-74 (20 October 1999), section 434 directs that NASA:

Shall establish and publish a price policy designed to eliminate price uncertainty for those planning to utilize the International Space Station and its related facilities for United States commercial use.

Further is stipulated that receipts collected by NASA:

Shall first be used to offset any costs incurred by NASA in support of the United States commercial use of the International Space Station. Any receipts collected in excess of the costs identified pursuant to the previous sentence may be retained by NASA for use without fiscal year limitation in promoting the commercial use of the International Space Station.

It has to be mentioned that NASA is allowed to waive part, or all of the marginal costs in the short term, in order to stimulate private investment during the formative period of business development. In response to this, NASA has developed a pricing policy [26] which is based upon the previously mentioned principle of allocating 30% of the internal and also 25% of the external resources to commercial users. The units are calculated on the basis of ISPR (International Standard Payload rack) units as follows:

Internal commercial sites = (27 ISPR sites) x (0.30 allocation) = 8

and for External Express adapters as follows:

External commercial site = (29 attached sites) x (0.25 allocation) = 7

Prices are established for two reference bundles as per table 8.13.

Resource	Internal Bundle	External Bundle
Location	One ISPR Site	One Express adapter
Energy	2.88 kWh	2.88 kWh
Crew-time (IVA)	86 hrs	32 hrs
Data capacity (Space-to-Ground)	2.0 Terabits	2.6 Terabits

Table 8.13: NASA Standard Bundle definition

The standard price is set at 20.8 million $ per bundle. Premium prices will be added or reduced according to the specific needs (e.g. 15,000 $ per crew-hour). The implementation of costwaivers (possible for scientific and educational projects and excluded for entertainment and advertising) and the procedures to process the proposals are outlined in [26]. Interesting in this context is the fact that NASA has decided to comply with ISO-9000 standards for this processing although the "entrepreneurial" principle for the criteria is clear:

Commercial enterprises, which demonstrate the highest private investment levels and involve products or services offered to non-government markets, will advance most quickly in the queue

The Canadian Space Agency makes a distinction between the early phase and advanced operations [11]. It is the intention to license both phases to a private company which will be charged the net cost price in the early phase, namely:

- L = Locker Site cost = 650,000 $ per three-month period.

- R = Resources:
    - Crew time = 15,000 $ per hour crew activity
    - Energy = 2,000 $ per ISS kWh
    - Communications = 100,000 $ per minute transponder time.

- T = Transportation cost (basically 44,000 $, unless other arrangements).

- S = Other services, not included in the Locker cost.

For the second phase, the payments to CSA will include a return part, so that CIAC, the private company created for this purpose, then will pay to CSA:

Price = L + R + T + S + Royalties (or other fees).

8.7 PHYSICAL DISTRIBUTION

DATA DISTRIBUTION

Two aspects have to be considered:

1. The Datalink between the Station and the Ground

During the MIR era, with a reduced number of groundstations and Data Relay satellites because of financial constraints, the problem of a lack of interactive contact was particularly witnessed. In order to meet requests from the user community, NASA installed the so-called MIPS (Mir Interface to Payload Systems) system on board of MIR. Using a dedicated Payload and Crew support computer this allowed the ESA astronauts on board MIR to store data in a buffer which was then successfully dispatched to the ground (and vice versa) outside of working hours, i.e. during passages when the crew was asleep [27]. Of course, this was a successful but still only a remedial step. For proper on-line data transfer a near-constant datalink is required as well as a higher data rate.

2. The distribution on ground to the end users

Until recently the world of mission control was a closed environment. All data were "controlled" and collected in one central place, the Mission Control Centre or the Payload Operations Control Centre depending on the respective application cases. However, as discussed earlier, for long-duration missions the scientist prefers to stay at his homebase and receive the data there. This was difficult in the past because this required costly installations and operations, but the Internet and its technologies, especially the WWW have changed this situation considerably last decennium. ESA developed and maintained the Interconnection Ground Subnetwork (IGS) testbed to interconnect all European sites for operations and experiment support and to provide operational links and services to NASA. However, it can be expected that the rapid evolution in the WWW environment will lead to further developments of this communication infrastructure in upcoming years.

CONTACT WITH THE USERS

ESA has decided to have a decentralised management of Payload operations for Columbus by establishing a number of so-called User Support Operating

Centres (USOCs). Under the central coordination of the Columbus Control Centres such USOCs will serve, depending on their level of involvement, as a Facility Responsible Centre, Facility Support Centre or Experiment Support Centre [28]. During the preparatory phase they will be the main contact for the users, providing them with advice and support. During operations, they will act as an interface between the ESA payload operations on board the Space Station and to the scientific user groups assigned to them.

The selection of USOCs follows dual criteria. A number of them are selected on the basis of specific technical experience to the point of being responsible for one ESA facility. On the other hand a geographical distribution is maintained to facilitate contacts with the users. The network of USOCs will no doubt evolve during the course of Station operations and presently comprises:

- CADMOS (F)
- MARS (I)
- MUSC (D)
- ESTEC (Nl)
- DAMEC (Dk)
- DUC (Nl)
- BUSOC (B)
- INTA/Inst. De Riva (E).

The International Space Station User Centre that is located in ESTEC, in the Netherlands takes a specific role. The Centre is equipped with following facilities:

- A High-Bay, housing a high-fidelity mock-up of Columbus
- A Virtual Reality/Multimedia Library
- A TV Broadcasting studio.

The objective is to provide a User Information Centre for all potential users, especially those less familiar with the space environment and procedures. Online information on the activities is provided under

www.estec.esa.nl/spaceflight/usercentre/.

Specific symposia are organised on the subject In order to facilitate contacts. The "First European Symposium on the Utilisation of the International

Space Station" took place in Darmstadt in 1996 [29] and a second one in Noordwijk, which was attended by more than 500 participants, took place in 1998 [30]. The aim of these symposia is described as [31]:

- To inform the User Community on the major progress made in the overall ISS Programme and to obtain overviews from our ISS partners.
- To inform in particular on utilisation possibilities for European users
- And to learn about the ideas and plans on how the European users wish to use these opportunities. This information will better help to learn about the user wishes and to orient the European research plans to make the best possible use of ISS.

It is of no surprise that after these two-way communication events, a considerable number of new proposals are submitted.

For the formal distribution of proposals, ESA uses Internet facilities to large extent. The so-called "Announcements of Opportunity" invite the users, grouped by discipline, to come forward with proposals for review. They can be accessed, in addition to other valuable information on European participation in ISS, via

www.estec.esa.nl/spaceflight/

INTELLECTUAL PROPERTY

We have to mention in this context the very important factor of the distribution and use of results from a legal point of view. NASA has settled this matter for its users in principle and prepared a first ISS Intellectual property reference guide which is available on the web site

http://commercial.nasa.gov/

The background and underlying principles have been explained in [32]:

- Overall, the Inter-Governmental Agreement (IGA) has acknowledged the existence of proprietary or commercially sensitive information and authorises for the establishment of strict procedures as protection
- Confidential data will be properly marked and each partner has agreed to respect the proprietary rights in, and the confidentiality of, properly identified and appropriately marked goods and rights

- For purposes of determining the country of inventorship, a territorial approach based on the ownership/registry of elements has been agreed upon. (So, if an invention is made in the Japanese laboratory, the patent must be filed initially in Japan).

In Europe this point is raising still considerable concern with industry as already mentioned in an earlier reference. Eurospace has worded European industry concerns [19] p. xi, as follows:

> *Space industry must obtain guarantees concerning:*
> - *The ownership of industrial property rights for any findings resulting from such experiment aboard the ISS, whatever the location of the experiment on board*
> - *The protection of any previous or background expertise even when disclosed for reasons of safety or management of the experiments to ISS partners.*

Palous et al. [33] feel that Industry would be a better "vehicle" with which to get in touch with potential customers. Contrary to Public authorities, industry has the right contacts (also in non-space sectors) and knows of interests for certain products in associated sectors. It has to be admitted that, with current globalisation efforts, there is some realism in this statement. An example in Europe is the application of fuel cells in cars which took place after Dornier was taken over by Daimler-Benz.

8.8   PROMOTION OF ISS

One could once again make a differentiation between two groups who have different interests and therefore need a different approach, namely the potential users of the space station and the general public. As far as the general public is concerned promotion of the ISS has some objective advantages compared to some other projects, such as:

- There is a lot of information on present space stations such as MIR, thus one can easily relate to this
- ISS is "personalised" by astronauts, with whom the general public can identify itself with
- The Space Station has a political dimension, which interests the public
- There are various ground facilities, such as mock-ups and trainers, which can be visited

- Last but not least, the Space Station is seen by the general public as a next step in manned space exploration

A more difficult task is to convince potential industrial users and promote ISS to this "client". The media mentioned under Physical distribution are a primary promotion tool for the user community, but risk remaining too restricted to such a semi-closed community. However, ESA has realised that a proactive approach to a broader "market" would require a business-oriented marketing specialist. Therefore, a contract is envisaged with a commercial agent (or a number of agents) having as their specific role:

- Promotion and marketing of ISS
- Selection of commercial customers
- Allocation of resources and operations services to customers
- Setting price structures in coordination with the Agency
- Handling payments for customers.

SPONSORING APPROACH

Although this can generate considerable income and therefore can be considered under the Price factor, using ISS as a sponsoring carrier would represent an important promotional tool.

Evidently, the income effect cannot be neglected; frustrated by the ever repeated questioning of the Space Station budget, a NASA official once uttered:

> **Maybe it would be easier to convince Bill (Gates) than to convince Bill (Clinton)...**

Tierney's ideas [34] for a sponsoring approach for Mars could be equally implemented for ISS, for example:

- Corporate Sponsoring: One could imagine that hi-tech companies would be willing to spend annual budgets in the order of 100 million $ as a "Space Station-sponsor", analogous to an Olympic games sponsors.

- Marketing tie-ins: Models of the ISS, T-shirts, and so on are comparable to the Mars estimates to be a market of minimum 1 billion $ yearly

- TV and Film rights: Exclusivity rights will be a touchy subject, but special events could be well covered under an exclusive agreement. The Olympic Games equivalent, collecting some 2 billion $ marketing/TV income for a 2 weeks event seems a valid reference.

Some recent examples related to this are:

- The Krunishev Space Centre obtained 4% of shares in the IRIDIUM network together with exclusivity rights and offsetting this by launch costs
- The Russian Space Agency (RKA – now RAKA) founded in 1997 the "RKA Centre of Marketing" for the ISS (min. price 10.000 $/m^2, real prices unknown) [35].
- Russian cosmonauts made publicity for Fisher Space Pens [36] and even for an Israeli brand of milk called Tnuva [37].

As previously mentioned, promotion budgets of large multinationals reach 3 billion $ per year; 10% of this budget from one single sponsor could cover Europe's yearly ISS exploitation costs. As an illustrative example, the Ferrari Formula 1 team collects 115 million $ (in 1998) from a limited number of sponsors who are selected from a list of potential candidates. Marlboro, who sponsors the team even without any relation to the product itself, provides 60 million $ as a main sponsor alone [38].

NASA is constrained at this moment by a number of legislative regulations which do not allow licensing fees to flow back to NASA. There are some examples in the merchandising area (such as the "Mattel Hot Wheels Mars Rover", a toy version of the Mars Pathfinder) but steps to fully explore the ISS potential in this field are still emerging. The previously described "Dreamtime" approach indicates a change in this attitude by opening the door for advertising. One of the preparatory initiatives was a one-day seminar at NASA's JPL in September 1998 on "Playing Among the Planets 98" where toy manufacturers, entertainment industry executives and others interested in space programme licensing were invited to attend.

Entertainment and Recreation

Even if this aspect is a commercial one, the promotional dimension should not be ignored. The IMAX film "Mission to Mir" (1997) shows the joint activities of NASA and Russian astronauts on board of Mir, taken by IMAX cameras. It also gave an insight into the Russian space activities (training in Star City, launch in Baikonour) and raises considerable public interest this way. A similar effect is evident in space theme parks that are equipped with physical or virtual reality models of ISS. In a similar category we can place the U.S. Space Camps which have a higher educational approach (but still had "graduated" in total more than 300,000 trainees by the end of 1999). Future ISS based projects include inter alia:

- **S*T*A*R*S,** educational opportunities to conduct research projects on board (a project offered by Spacehab) that intends to make use of Spacehab's "Enterprise" module attached to ISS. This project will be the continuation of the present STARS programme aboard the Shuttle, involving some 3,500 schools mainly in Australia, China, Israel, Japan, Singapore and the U.S. The schools develop a number of research projects (biosphere study, effect of microgravity on animals and plants in space, effect of microgravity on chemical-fibre growth) and pay Spacehab 60,000 $ collected mainly from local sponsorship for the execution of such experiment. As B. Harris, Spacehab's chief scientist expresses [39]:

   This is not a program that we expect to make a lot of money on, but it gives us an opportunity to get the word out about space and space commerce.

- **"Lecture from Space",** planning to have on board astronauts giving 10-15 minutes lectures from Corporate sponsored financing. If access via the Internet is provided as planned to the Station, it is evident that activities of this kind will grow exponentially.

Probably it is impossible to attribute the ISS support in the U.S. to a single one of these aspects. Still, a 2000 survey showed an even increasingly favourable opinion in the U.S. (compared to the 1999 results) for space exploration in general and the International Space Station in particular (even if progress between the two surveys has been rather disappointing). Even an increased willingness to bring the U.S. space budget back to 1 percent of the federal budget was noted, as can be shown in figure 8.4 [40].

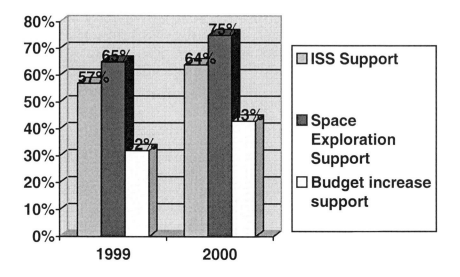

Fig. 8.4: Increase in Space Support

8.9 PHILOSOPHY AND ISS

The brief historical overview, as represented in table 8.1, indicates a number of important facts.

- Initially the rationale for a space station was the need for interplanetary travel as a permanent "resupply base", mainly for fuel
- This is linked to the desire of mankind to explore new frontiers
- Later, mostly U.S. politicians added the political dimension. They saw the station as a good chance to link their "allies", under the key objective of the U.S. civil programme promoting international space cooperation.
- Afterwards, the decision to add Russia was a symbol of the end of the Cold War combined with the U.S. interest of encouraging Russia to conclude agreements to stop the proliferation of ballistic missile technology as well as to support Russia economically and politically.

The official text of NASA is shown in table 8.14 [41].

> **WHY A SPACE STATION?**
>
> The Mission of the International Space Station is to enable long-term exploration of space and provide benefits to people on Earth.
>
> - To create a permanent orbiting science institute in space capable of performing long-duration research in the materials and life sciences areas in a nearly gravity-free environment
> - To conduct medical research in space
> - To develop new materials and processes in collaboration with industry
> - To accelerate breakthroughs in technology and engineering that will have immediate, practical applications for life on Earth - and will create jobs and economic opportunities today and in the decades to come.
> - To maintain U.S. leadership in space and in global competitiveness, and to serve as a driving force for emerging technologies.
> - To forge new partnerships with the nations of the world.
> - To inspire our children, foster the next generation of scientists, engineers, and entrepreneurs, and satisfy humanity's ancient need to explore and achieve.
> - To invest for today and tomorrow. Every dollar spent on space programmes at least 2$ in direct and indirect benefits.
> - To sustain and strengthen the United States' strongest export sector – aerospace technology- which in 1995 exceeded $33 billion.

Table 8.14: NASA's Space Station Manifest

Still, we can note differences when comparing NASA's with ESA's "manifest for the participation in the Space Station", which defines:

Utilisation Benefits
- Scientific research in a physical environment not possible on Earth
- Observation and study of the Earth and the Universe from a vantage position outside the Earth's atmosphere
- Technology innovation and development of new applications.

Build-up of Know-How
- Development of the key elements required for the operation of a space station without the need for Europe to develop the complete required space and ground infrastructure by its own means.

Political Benefits
- Fostering of international cooperation including in particular the integration of Russia into cooperative structures
- Preparations for Europe's place in global cooperation structures of the future.

We can note differences in the leadership and financial motivation, but it is interesting to note the emphasis on the visionary aspect from all sides. Space activities were initially put under the scientific scope and later extended to technological aspects. However, the philosophical dimension has never been so clearly underlined as under the ISS era.

In this context, we should also not overlook the so-called "Sagan argument". Carl Sagan has always propagated space exploration to be a "world-wide endeavour". His reasoning is based on a technical and a philosophical dimension.

- **Technical**: As a result of historical technical cultural differences, engineers in different countries have developed different skills in various specialities. By joining these skills, no nation needs to acquire such skills at a very high cost. (E.g. MIR mechanical experience combined with the European data management experience of DMS-R).
- **Philosophical:** Outer space belongs to no-one, so competition makes no sense, cooperation does.

During the Apollo times, 2 million people were directly or indirectly involved as well as thousands of dispersed manufacturers. All had to complete parts to a standardised and high precision within a very strict deadline. The experience gained from this has led to a considerable spin-off. If we extrapolate this to the ISS era, this project will involve people from 16 nations, with different technical cultures. They will have to make parts that fit together and undoubtedly they will learn from each other; this will certainly lead to a broader spin-off.

The deeper philosophical dimension of equality is reflected in [42]:

In exploring the space frontier it is important that we as a species come to understand ourselves and that we finally become equal, regardless of gender, race, color, age and origin, and it may be from this philosophical basis that we explore. The international space station is probably humanity's most important program as the next logical step in the exploration of space. Not only because it is a platform on which scientific experiments will be conducted, but also because either by design or by default, it will become a great social experiment. It will become an international platform in space where for the first time people of different cultures, races and genders will repeatedly assemble to work together. These crew members will have to learn to overcome their cultural, traditional, racial and gender divides, and for this reason the ISS may be the springboard from which to go forward as humans exploring space.

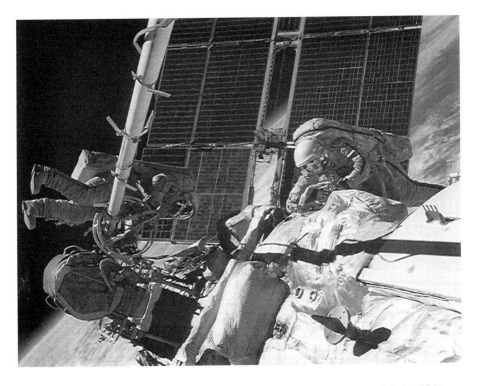

(photo: ESA)

Fig. 8.5: Russian Cosmonaut Y. Gidzenko and ESA astronaut T. Reiter working together in space

8.10 CONCLUSION

With the International Space Station, the international community will have access to a unique laboratory in orbit superior to all preceding ones in terms of features, capacity and logistics. It will provide a number of research opportunities as well as an unprecedented number of industrial potentials.

In the past, often the Life Cycle Cost concept of big programmes has been neglected. European industry, supported by industrial policy considerations, has been more interested in the construction of new projects rather than operating existing ones. For ISS new approaches are envisaged in Europe. On the one hand a better role for industry during the exploitation phase is envisaged, on the other hand the search for commercial users.

ISS provides a large stage for industrial applications. These range from technology demonstrations for certain applications which could operate eventually in an automatic mode, but initially will benefit from human presence during the validation phase, to applications in different disciplines. Thanks to regular telemetry and communications, possibilities to change, adapt or modify experiments and the availability of trained operators permanently on board we can only imagine the industrial possibilities. It is difficult to forecast what will evolve over the next few years, as new experiments build further upon the results of previous ones. The commercial potential is acknowledged by industry but a number of steps still need to be taken before a full development takes place. First of all, a number of question marks such as the pricing policy need to be settled. Furthermore a considerable number of legal aspects will equally need to provide acceptable modalities. This is in particular the case with intellectual property regulations, in view of the lack of proper regulatory guidelines. Only once these prerequisites are satisfactorily solved can it be expected that a deeper involvement of industry, in the form of industry originated proposals, will materialise. From that point onwards, we may witness how industry gradually takes over the operational cost of a publicly financed project.

Failing to come to such an arrangement would be regrettable. ISS is an important social experiment, where people of different cultures and also political ones, will live, build and work together on an equal basis and on a scale larger than ever before. This philosophical dimension should not be forgotten and we would miss an enormous chance for humanity if this would lead indirectly to an improvement in intercultural relations.

REFERENCES CHAPTER 8

1. NASA, *Space Station Freedom Media Handbook*, (May 1992).
2. SMITH, M., *Space Stations*, order code IB93017, Congressional Research Service. (1995).
3. MESSERSCHMID, E. and BERTRAND, R.: *Space Stations*, (Springer, Berlin 1999).
4. ROTHENBERG, J. and NICOGOSSIAN, A., *Commercial Development Plan for the International Space Station*, (NASA, 16 November 1998).
5. SWALES AEROSPACE: *Options for Managing Space Station Utilization*, (NASA, October 1999).
6. FERSTER, W., NASA Works To Counter Anti-Commercial Culture, *Space News*, (August 9, 1999) p. 4.
7. DICKEY, B., Everything Must Go!, *GovExec.com*, (August 1999).
8. GOLDIN, D., The National Importance of the Development of Space, *Speech for the US Chamber of Commerce*, (Washington, 16 March 1999).
9. TUMLINSON, M.: A Lack of Vision, *Space News*, (Oct 5, 1998), p.5.
10. KPMG, NASA, *Commerce and the International Space Station*, (NASA study, November 1999).
11. CORBIN, S., ISS Commercialization : Canada's Path Forward. *Paper presented at the Project 2001 Workshop.* (Berlin, June 2000).
12. TAILHADES, J., Industrial approach to ISS Utilisation for technology, *Proc. on Symposium for Space Station Utilisation,* ESA SP-385 (ESA, Noordwijk, December 1996) pp. 463-468.
13. UN, *Treaty on Principles Governing the Activities of States in the Exploration and Use of Outer Space, including the Moon and other Celestial Bodies*, (UN, London/Moscow/Washington, 10 October 1967).
14. FARAND, A., Space Station Cooperation: Legal Arrangements, in *Outlook on Space Law over the Next 30 years*, (Kluwer Law International, 1997), p.125-133.
15. LARSON, W. and PRANKE, L., *Human Spaceflight*, (McGraw Hill, New York, 1999).
16. TUMLINSON, M., NASA Shouldn't be a Landlord, *The Wall Street Journal*, (April 8, 1997).
17. FARAND, A., Legal aspects of International Space Station utilisation: a European perspective. *Paper presented at Project 2001 Workshop*, (Berlin, June 2000).
18. ESA, *Exploiting the International Space Station – A Mission for Europe*, ESA-BR-141, (ESA, Noordwijk, February 1999).
19. EUROSPACE, *Industrial Utilisation of the ISS by European Industry*, annex to ESA-BR-141, (February 1999).
20. BEARDSLEY, T., Science in the Sky, *Scientific American*, (June 1996) pp. 36-41
21. MOTT, M., New Markets: The Role of Exploration, in HASKELL, G. and RYCROFT, M., *New Space Markets* (Kluwer, Dordrecht, 1998), pp.86.
22. BERGER, B., First Commercial Deal Penned for Station, *Space News*, (December 20, 1999) p.3.
23. NASA, *International Space Station Fact Book*, under http://spaceflight.nasa.gov (July 1999).
24. HARR, M. and KOHLI, R., *Commercial Utilization of Space*, (Batelle, Columbus, 1989).
25. FLAHERTY, C., Pricing Policy, Structure and Schedule for US Resources and Accommodations on the International Space Station. *Paper presented at STAIF-2000* (Albuquerque, February 2000)
26. UHRAN, M., Economic Development in the International Space Station. *Paper presented at STAIF-2000* (Albuquerque, February 2000).

27. BESSONE, L., DE JONG, F. & NESPOLI, P.: Crew Support Tools for Euromir 95, *ESA Bulletin*, 88 (ESA, Noordwijk, November 1996).
28. REIBALDI, g. et al, The Microgravity Facilities for Columbus Programme, *ESA Bulletin* 90, (May 1997). pp. 6-20.
29. ESA, *Proceedings of the Symposium for Space Station Utilisation*, ESA SP-385 (ESA, Noordwijk, December 1996).
30. ESA, *Proceedings of the 2nd Symposium for Space Station Utilisation*, ESA SP-433 (ESA, Noordwijk, February 1999).
31. FEUSTEL-BÜECHL, J., Objectives of symposium and European Programme Review. *Proceedings of the 2nd Symposium for Space Station Utilisation*, ESA SP-433 (ESA, Noordwijk, February 1999). p.7.
32. BROADWELL, M., Intellectual Property and the Economic Development of the International Space Station, *Paper presented at STAIF-2000* (Albuquerque, February 2000).
33. PALOUS, J. et al., An Industrial approach for ISS Utilisation *Proc. of the 2nd Symposium for Space Station Utilisation,* ESA SP-433 (ESA, Noordwijk, February 1999) pp.113-116.
34. TIERNEY, J., How to get to Mars (And Make Millions !), *The New York Times Magazine*, (May 26, 1996) pp. 21-25
35. KOMMERSANT DAILY (translated from Russian): Establishment of the Advertisement Space Agency, *Kommersant Daily*, (Feb. 14, 1998).
36. FLORIDA TODAY, *Mir Cosmonauts Pitch Pens on U.S. Television*, (8 February 1998).
37. FLORIDA TODAY, *Russian cosmonaut films milk ad on MIR space station*, (21 August 1997).
38. FOCUS, Formel 1, *Focus*, (3/1999) p.150.
39. BERGER, B., Program Puts Children's' Project Aboard Shuttle, *Space News*, (April 10, 2000), p.20.
40. BERGER, B. and MAGNUSON, S., NASA, Exploration Score High in Public Opinion Poll, *Space News*, (April 17, 2000) p. 14.
41. NASA, *Why a Space Station?* http://spaceflight.nasa.gov/station/reference/factbook/why.html (update 8 March 2000).
42. KLEINBERGER, R., Why Explore Space? *Space News*, (March, 6, 2000).

Chapter 9

CASE STUDY: SPACE TOURISM

9.1 INTRODUCTION

One of the big advocates of space tourism, Ashford, once noted:

> *Space Tourism will begin 10 years after people stop laughing at the concept. Recently people have stopped laughing…*

Previous approaches mainly focused on the "how" of Space Tourism. However, as there is no reasonable concept possible with the present upload costs, they immediately failed any economical evaluation and, to a certain extent, at the same time condemned Space Tourism to the Science Fiction World.

We have to add to this a resistance to big exploratory changes, as described in a previous chapter, and unavoidably we find a number of eminent people claiming that Space Tourism is utterly impossible. (In line with their ancestors claiming a similar thing, with comparable arguments on the discovery of America, travelling by train, Iron-clad ships, Airplanes, Rockets and so on).

However, from a marketing point of view, we may look differently at this topic. If there is a market, history as shown that one day a product will be conceived which satisfies this market. From the tourism perspective, there is a clear request for novel products because the traditional markets are reaching maturity in their Product Life Cycle. From the potential customers side, early research confirms the demand exists for such market as well, even if initially high prices and high risks are involved (both of them being probably the ultimate stimulus for the first tourists).

Therefore, the classical marketing approach will be used hereafter, less focusing on the product itself but more on the boundary conditions to which the product will have to comply.

9.2 GLOBAL BOUNDARY CONDITIONS

Public access to space

Instead of the more popular term "Space Tourism" it would be better to talk about "Public access to Space". Indeed the key element is to provide systems which are safe and comfortable enough to transport paying customers, from the general public into space.

There will of course always be medical and psychological prerequisites, but they shall be considered to be in the same order of severity as for example participation in Antarctic or mountaineering expeditions presently. A considerable step in this direction was the reflight of J. Glenn in 1998 on board the U.S. Space Shuttle. Although undoubtedly Senator Glenn had maintained a good basic condition (he was still flying airplanes intensively), his age of 76 years clearly showed that a wider percentile of the population is basically fit enough to make a space trip.

Also many scientists and astronauts have contradicted the popular belief that travelling into space will be spoilt by space sickness. Space sickness is just a form of motion-sickness that can be prevented by using normal travel-sickness medicine. From this point of view, medical and physical prerequisites will not put a constraint on public access to space, particularly when taking into account the fact that demand will be, probably for a considerable time, higher than the supply of opportunities.

Legal context

The term "Space" cannot be physically defined exactly because there is a gradual transition. The legal definition of "Outer Space" is still under definition. What is generally applied is a functional limit, i.e. air space extends where planes can fly. This height is arbitrarily put at 100 km above the Earth (on the other hand NASA maintains a 50 miles border for the "astronaut status" designation). To illustrate the consequences of this lack of clarity: the country of Colombia claims that their jurisdiction extends to much higher altitudes and they would like to get royalties for the passage of communication satellites through "their" territory.

States have jurisdiction in the air space above their territory. For outer space they can only assert jurisdiction on specific space objects registered by them, on the basis of international agreements such as the UN Registration

Convention. Moreover, the UN Outer Space Treaty of 1967 [1] prohibits national appropriation of outer space or parts of outer space. Also this is leading to vigorous discussions for example on the registration of land on the Moon and on Mars, as are happening now. This aspect is evidently important for the subject of Public access to Space (property rights, liability, insurance, commercial law) and will require further legal definition [2], a task which is now mainly coordinated by the legal subcommittee of the UN COPUOS (Committee on the Peaceful Utilisation of Outer Space).

Tourism

Tourism is still a growing market, even if the growth is slightly less than the yearly growth forecasts of the early 1980's, it still reached some 4.3% growth between 1989 and 1998.

In the 1960's and 70's we talk about a boom in tourism. Economical growth, availability of cars and, specifically, the access to gradually cheaper flight tickets gave raise to mass-tourism. With the charter flight, ever further-away destinations came within reach of the consumer satisfying the "competitive" motivation of the traveller (showing what one could afford, being able to talk about places others have not visited). Tourism marketers were then mainly emphasising their marketing efforts on the Price and Promotion aspects, the Product and the quality thereof, has often been neglected.

This could be one of the reasons why in the 1990's the growth rate suddenly reduced. Indeed, this was not only due to a number of "environmental" aspects such as economic problems, fear of unemployment and even fear of terrorism, but also to the continued use of the old product mix. Marketers realised that in the tourism market also the Life Cycle effect took place; after the mass-tourism and "Package-tourism" (club formulas, organised tours) of the 1970's and the 1980's this effect started to show a "burn-out" effect. It was realised that the existing Product Life Cycle was levelling, whereas contrary to other consumer goods, no new PLC had been initiated as yet [3].

One of the often-quoted problems is the difficulty of researching new tourism products. Ilsley [4] admits that forecasting models are difficult to develop in the tourism sector because the purchase of the Product is sporadic and it needs a long lead time. Furthermore, there is a very big difference between the intention-to-procure and the factual procurement, so that market prognosis can be very misleading. Nevertheless, the author still strongly advises using the techniques of New Product development in the stagnating market. Stipanik [5] stresses the point that new technologies are not considered fully in the New Product Development for Tourism. Although he emphasises aspects such as virtual reality and artificial tropical environments, his basic remark on the involvement of new technology areas is very applicable to our specific "Space Tourism" case as a potential impetus in the Tourism Life Cycle.

9.3 DISTINCTIONS IN SPACE TOURISM APPROACHES

In general we distinguish between 4 types of activities:

Orbital flights using existing technology

The idea is that existing systems, such as the Space Shuttle, are slightly modified and will transport people "out in space". Such step is comparable with the first public flying experiences in the 1920's, when First World War bombers were converted to carry a few passengers. These flights, such as the London-Paris Express of 1919 (using a De Havilland DH16) were very uncomfortable and certainly not without risk and yet they attracted a certain population due to the adventurous character of it. On top of this, many commercial airlines have their roots in this period, as a result of the surplus of planes and experienced pilots. One example is Pan Am, which started operations in 1923 with nine ex-Navy aquaplanes (saved from a scrap yard).

Suborbital and orbital short duration flights

The key to this could be the X Prize Foundation. In analogy with the 25000 $ Orteig Prize for the first successful crossing by an airplane over the Atlantic and claimed by C. Lindbergh in 1927, the X Prize Foundation – a nonprofit, educational group - has launched a competition. 0n May 18th, 1996 a 10 million $ prize was issued with following boundary conditions [6]:

- The spaceship must be privately financed and built. No government projects are allowed
- It must carry at least three adults to at least, a 100-km altitude and return them in good shape

- It must be reusable and able to fly twice in 14 days; the craft must return to Earth in a reusable condition.

Clearly the aim of the organisation is to drive designers in the direction of an affordable public space transportation system, challenging "present notions that only governments can send people and cargo into space". Therefore the Price Competition aspect is only one, be it very visible, aspect of the Foundation. P. Diamandis, the driving force behind this initiative, describes the Foundation's mission statement as follows [7]:

- Organising and implementing competitions to accelerate the development of low-cost spaceships for travel, tourism and commerce
- Creating programs, which allow the public to understand the benefits of low-cost space travel
- Providing the public with the opportunity to experience the adventure of space travel directly.

Under the category of suborbital transport we can also classify the so-called "Boost-gliders" which combine space tourism with commercial transport. A Boost-glider takes off at a commercial airport, climbs out of atmosphere (the boost phase) reaching near-satellite speeds, and then returns to its destination using centrifugal force (the gliding phase). The duration of an antipodal flight (say Paris-Sidney) would then be in the order of 75 minutes. As this represents a considerable saving in time compared with the present 20 hours travel time, there are certainly market prospects in this field.

Space Hotels

Before any Space Hotel project can be realised, the transport cost problem needs to be solved first, in line with the previous two categories. The eldest concept that was technically detailed by the Russian "Space Pioneer" Konstantin Tsiolkovsky dates from 1929 [8]. He made a proposal for a 3000 m long cylindrical habitation module in space with a diameter of 3 m providing place for 300 families. It is extraordinary to read how this visionary scientist considered various aspects such as the rotation (to provide artificial gravity) and "gardens" in the middle serving a closed loop life support system.

Wernher Von Braun, who conceived circular towns, developed a later and more realistic model in 1952. The rings were rotating in order to provide artificial gravity at the outside perimeters.

Even if Space Hotels may appeal most to the imagination, the economic and technical feasibility do not allow their realisation in the next few decades.

Fig. 9.1: Space habitat proposal from W. von Braun (1952)

Tourism on Moon or Mars

These two places are mentioned on purpose. After man had stepped on the Moon, it is evident that many people dream of doing the same. Looking to Earth from the Moon is a dream of mankind. This enthusiasm can be illustrated by the "Moon Register" which the Travel Agency Thomas Cook started in 1954; also PanAm had a sort of "waiting list" (the First Moon Flights Club) which ran from 1964 till 1971 and had some 100,000 people from 90 countries registered.

The technical interest in Mars can be best illustrated using table 9.1, where the main characteristics of Mars and the Earth are compared.

	MARS	EARTH
Equatorial radius	3397 km	6378 km
Axial Rotation ("day")	24 hrs 37 min 23 s	23 hrs 56 min 4 s
Rotation around sun	687 days	365 days
Inclination	23.98 degree	23.45 degree
Gravity	3.7 m/s^2	9.8 m/s^2
Atmospheric pressure	0.01 bar	1 bar
Temperature (Equator)	-50 °C to +20 °C	15 to 30 °C

Table 9.1: Mars/Earth Characteristics comparison

One soon notes a number of remarkable equivalencies. Both the axial rotation (important for circadian rhythm) as well as the inclination to the Sun (seasonal changes) are comparable. On the other hand, gravity can be considered as more comfortable than on Earth. Even the temperatures (comparable with Antarctic ones) and very low atmospheric pressure are no major "show stoppers" for manned activities on Mars. Another, important factor, are the presence of CO_2 and H_2O (certainly at the poles and most probably under the surface) and other "building materials" very familiar to us. This has led to various theories and approaches on "Terraforming", by which is meant the gradual changing the conditions on Mars (mainly by creating an atmosphere) to allow a living environment comparable with the one on Earth (be it with less gravity and perhaps even more comfortable). An excellent overview and a convincing testimony for further Mars exploration can be found in Zubrin [9]. In view of the travel time involved of, realistically some one thousand days, it looks evident that Mars falls into the category of "Space Colonialisation" rather than Space Tourism.

9.4 ANALYSIS OF THE MARKET

The keyword in this respect is "Adventure Tourism"; an exponentially developing market. Tourist operators are constantly looking for new products to satisfy a market which demands "exclusive" products. In the 1980's tourism experts were convinced [10] p.384 that:

A passenger module will be developed for the space shuttle that will carry passengers (...) or act as a hotel module itself.

Probably the only reason why this market has not developed is the high investment costs and the reluctance to share publicly funded space transportation services with private investors. It can be assumed that private initiatives (such as the previously mentioned X Prize) or new - PPP based - cooperative programmes will change this situation.

The yearly market volume for Tourism is presently in the order of 3400 billion $, which means that 1% of this market would be the equivalent of the present yearly public space expenditure. In order to evaluate the importance of tourism we should not forget that tourism generates 10.2 % of the world gross national product, and accounts for 10.9 % of all consumers spending. From these figures (source: World Tourism Organisation, 1999) it means that tourism is the world's largest industry in terms of gross output providing a world-wide employment to 200 million people (i.e. 8% of total world employment).

Presently, some 110 billion $ are spent on adventure recreation in the U.S. alone, with some examples quoted [11] as:

- 20,000 $ to sail to the Arctic
- 70,000 $ to climb Mount Everest
- 150,000 $ for a sight seeing tour around the world aboard a small executive jet.

Naisbitt [12] p.163 uses the following definition for adventure recreation:

> *A variety of self-initiated activities utilizing an interaction with the natural environment that contain elements of real or apparent danger, in which the outcome, while uncertain, can be influenced by the participant and circumstance.*

He points out that e.g. bungee-jumping, non-existent 25 years ago, now accounts for some 100 million $ annual turnover in the U.S. alone.

Also Buzz Aldrin who was one of the first people to walk on the Moon on 16 July 1969 is now a convinced advocate for Space Tourism and is advising various firms as a consultant. At a conference in June 1999 he declared:

> *Space Tourism is the logical destination, the next step for that market.*

Aldrin also maintains an Internet site on this subject under:

www.sharespace.org .

It is estimated that the Space Tourism Market will develop in several stages. In the pioneering phase of the first ten years a market of some 20 billion $ is expected. One of the first documented and practical business propositions was the Space Voyage project, developed by a Seattle based company specialising in exotic trips. The programme was designed to send up to twenty tourists using a single stage booster called Phoenix E. By 1987, 188 persons had made a 10% downpayment of 5,000 $, which clearly illustrates the market potential [13].

Market surveys suggest the following potentials:

- In Japan, it has been reported in 1994 that well over one million people per year would be willing to pay 10,000 $ for a short stay in space. The survey further came to the conclusion that about 70% of the Japanese were interested in a space trip [14]
- In a 1995 Canada/U.S. telephone survey 45.6% of the people contacted confirmed that they would pay three months salary for the trip; some 10.6% (scaling to 11 million people) would even be willing to pay one year's salary. Of interest for the growth potential is that the response was 3 times more favourable in the "under 40 years" category, compared to the "60-80 years" category. From an interest point of view, it can be estimated that in the U.S./Canada over 60% of the population are interested in a space trip [15]
- In 1997, a similar survey was performed in Germany and came to the conclusion that 4.3% of the Germans would be willing to spend an annual salary for a holiday trip to space, however, only 43% of the Germans were interested in such a trip [16]
- In 1999, the survey was repeated in the U.K. [17]. In this case, only 34.7% of the respondents expressed that they would like to undertake a space trip, thus considerably lower compared to Japan and North America. Of those, on the other hand, 11.6% of the respondents (equivalent to 3.5 million people) declared themselves willing to pay one-years income for such a trip
- International polls indicate a market of about 20 million persons per year, with space tickets in the order of 1000 $. But even at a space ticket of 500,000 $ for a Suborbital flight or 5-10 million $ for a full space

stay, it is estimated that there will be sufficient demand for each opportunity offered [11]
- Using sensitivity analysis, Abitzsch [18] grouped the various surveys and came to the conclusion that a "ticket" for a space trip of the order of 50,000 $ (1996 prices) would create sufficient interest. Provided the supply side can then satisfy the demand (which is not evident because no concrete developments are on their way presently), he comes to the conclusion that such a price may lead to a market of more than 60 billion $/year. Even if the assumptions are optimistic, they illustrate the now widely recognised potentialities of the space tourism market.

If we express price figures of this order (lower prices cannot be reached and higher ones seem to lead to a rapid drop in demand) and compare the previous surveys by expressing the number of people interested per country as a function of the proportion of income that they are willing to pay, the results are as per figure 9.2.

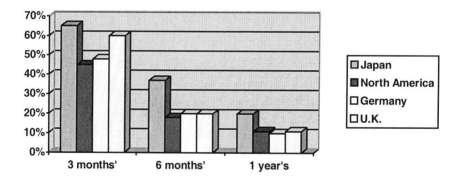

Fig. 9.2: Proportion of income willing to pay for a space trip per country

Putting these facts together, it is clear that there is a market demand and a market potential, the problem remains of the high investment cost and the associated risk. It will require an entrepreneurial risk, similar to Motorola's Iridium project, to start off the business. The billionaire Richard Branson (Virgin Group) established a company called "Virgin Galactic Airways" in April 1999 targeting a 200,000 people market willing to pay up to 100,000 $ for a trip. It is initiatives like this that will probably one day spark off Space Tourism [19]. Another billionaire and hotel tycoon, Robert Bigelow, intends to invest 500 million $ in his "Bigelow Aerospace" company.

Progress can be followed under

www.bigelowaerospace.com .

On a more institutional basis, NASA and STA, the Space Transportation Association, conducted a study through a cooperative agreement on the potential for future general public space travel [20]. The study resulted in recommendations in various technical as well as legislative fields and suggestions for redirection of certain NASA programmes such as the X-33 and X-34 experimental space transport systems. The study concludes:

> *We now see the opportunity of opening up space to the general public, a "sea-change" in our half-century sense that people in space would continue to be very few in number, would be limited to highly trained individuals who, at personal risk, would conduct mostly taxpayer supported scientific and technical activities there under government purview. Now the dream of very many of us during the Apollo era that we could someday take a trip to space for our personal reasons is finally approaching realisation.*

9.5 PRODUCTS OFFERED

PRODUCTS BASED UPON EXISTING HARDWARE

Already in 1979 plans were developed to place a module in the Shuttle Cargobay which would have had a capacity for 74 passengers. At that point in time the price tag was estimated at a - very considerable - 3.6 million $ per passenger, based upon a rate of twelve starts per year.

A first "Public access to Space" programme of NASA, unfortunately, coincided at the early phase with a dramatic event. Indeed, the first school-teacher to lift off to space, Christa McAuliffe, who was a part of this programme, was a member of the unfortunate crew who were killed in the Challenger accident on 28th Jan. 1986. It was the intention of NASA to take people from various professions on board of the Shuttle in order to propagate the "Public Access" idea. After the accident it became clear that the public "accepted" that professional astronauts ran these types of risks but were very upset about the loss of civilian life. This also teaches us that the reliability of launchers will have to be high enough to start Space

Tourism because an accident in the early phase may lead to violent public reactions.

Also in this category we can put a number of discussions, which took place in the framework of the Russian Mir missions. A Russian crew when launched with the highly reliable Soyuz rocket, consists of three persons. However, the Soyuz capsule only needs a maximum of two cosmonauts to be manoeuvred, so that a third "passenger" could be envisaged. Such passenger could then stay onboard during the hand-over phase (typically eight days) and then return with the previous MIR crew. It has to be noted in this context that the capsule landing on solid ground in Kazakstan is relatively demanding and puts considerable requirements on the physical condition of such a passenger.

In 1999, a private individual, Mr. Llewelyn tried to gather the money for such flight but failed; it cannot be excluded that another person, who comes up with the estimated 10-20 million $ "ticket" price will fly one day to the MIR Station.

The biggest chances may occur in the end phase of the current space transportation systems (i.e. without depreciation costs). It could be well be imagined that at the end of the operational lifetime of the Shuttle fleet a private initiative in this direction is developed. In the previously mentioned MIR Space Station such a phase is presently existing. After the Russian Government has stopped financing the station, a small Netherlands based and private company, called MirCorp is leasing the station from the Russian Company RSC Energia (the legal owner of the Station) [21]. The purpose of the company is to operate as a direct link between commercial users of Mir and the space station's Russian operators.

Besides such plans as using the station for broadcasting and filming purposes, the opportunity of a paid trip to the MIR station is one of the commercial possibilities envisaged. A first agreement of this kind was signed on 19 June 2000 between MirCorp and Dennis Tito, a former U.S. space programme engineer who founded Wilshire Associates; a California based investment management consulting company. The "ticket" for a seven to ten days trip to MIR, planned for early 2001, is said to cost Mr. Tito in the order of 20 million $.

His motivation to do such trip is indicative probably for what will be the first group of space tourists, as he stated [22]:

The launch of Sputnik in 1957 influenced me to become involved in the space program, and I was one of the first persons to graduate with a bachelor's degree in aeronautics and astronautics. As the result of MirCorp's own pioneering efforts, I will be able to accomplish something I have been wanting to do for more than 40 years.

Progress on this endeavour called the "Citizen Explorer Program" by MirCorp can be followed on the company's website under

www.mirstation.com

We have to mention here a number of on-going projects which, in a broader context, relate to this "Public access to Space":

- Around the Millennium event, the German Postal services picked out at random a number of greeting cards from the ones invited for that purpose and transported them on a probe into deep space
- The company "Encounter" plans a probe with coded messages (25$ per message) and the transport of human hairs (allowing aliens to decode the DNA) by 2001.
- A completely different approach was followed in Germany in December 1999. A Lottery was organised whereby the first prize was a "guaranteed" space trip. Perhaps also due to the timing around the Millennium, the lottery had a considerable success and at the same time brought Space Tourism closer to the general public.

Even if some of these projects seem a bit on the "opportunistic" side, they show clearly again the market potential for entrepreneurs with initiative in a market, which is very close to the imagination of the public.

SPACE TRANSPORTATION PROJECTS

The key to all Space Tourism issues is the transport costs. If we remain at the present transportation costs of some 20,000 $/kg, then we need for each passenger (including equipment and minimum necessities a total of 100 kg/person) "ticket prices" in the order of 2 million $. This will frighten off any potential investors and is a "no-go" situation. NASA bases replacement of the U.S. Shuttle on the X-33 reusable concept. This is a Vertical Take-off/Horizontal Landing vehicle which is presently undergoing testing and is intended to operate under the name RLV/Venturestar. The target is to bring the launch costs to approximately one tenth of the present ones.

Higher safety is reached using Horizontal Take-off/Horizontal Landing concepts. This is the basis of developments in Europe such as HOTOL and Saenger. In the medium term the purpose of these vehicles is suborbital transport and therefore seat capacities in the order of 100 seats are foreseen. Such concepts are studied in Europe by ESA in the so-called FESTIP programme (Future European Space Transportation Investigation Programme). Similar to this, NASA is working on the NASP project. In view of the development costs and lead times, probably these carriers will be preceded by simpler projects targeting the early Space Tourism market. Recent projects are briefly described hereafter:

Spacebus

A carrier plane, using a narrow body airliner-like fuselage (cf. Concorde) able to carry some 50 passengers. This design could use previous Concorde technology and has a number of conventional aspects such as a pilot cockpit and horizontal take-off and landing capability. The lift-off weight is in the order of a Boeing 747 and the direct operating costs at steady state, are estimated per flight in the order of 220,000 $ (bringing the cost per seat in the order of 4,500 $) [23].

Ascender

The design is an evolution of the X-15 design. It is a small aircraft (15 m, only 5 tonnes lift off weight) intended to be equipped with a crew of four (one pilot and three passengers). The plane takes off on a conventional runway and is under development by Bristol Spaceplanes Ltd. At 8 km height the rocket engines are activated to take the plane to a height of some 100 km where the passengers can witness a short zero-gravity stay. The plane lands after a 90 minutes trip and would be able to perform several trips per day. Using conventional technology, such as the engines, this plane could be developed for some 3.4 billion $ and operates at a cost of about 30,000 $ per flight, corresponding to a total cost per seat (including depreciation and overheads) in the order of 15,000 to 20,000 $ per passenger. [24].

Kanko-Maru

This Japanese project which is sponsored by the Japanese Rocket Society and the Kawasaki Concern follows a different approach. It is a reusable Single Stage To Orbit (SSTO) rocket vehicle with vertical take-off and landing. (Diameter 18 m and 23.5 m height with a reference mass of 61 Tonnes).

The vehicle could carry 50 passengers up to 200 km and is intended to make two earth orbits per flight. The design foresees large windows and a Microgravity "amusement room", hence a more attractive tourist service. The designers assume a ticket price of 90,000 $, based upon a daily trip and a 30 years lifetime (which are both rather optimistic assumptions). [25].

These examples already lead to a first observation: there are various initial steps which could lead to early space tourism but the higher the tourist comfort, the higher the development costs and the resulting ticket. This observation will most probably play a considerable role when a commercial choice has to be made.

THE SPACE HOTELS OFFERED

Shimizu Hotel

The design that is most often found in literature, is the concept of the Shimizu Corporation of 1989 [12]. It is a circular design (cf. the earlier mentioned von Braun project) with a 70 m radius. The rotation is foreseen at some 3 rpm, providing for an artificial gravity of 0.7g. In the outer ring, as shown in figure 9.3, 64 guestrooms are foreseen of 7 m in length and 4 m diameter.

The corporation is targeting to put the hotel in operation around 2020. Taking into account a forecasted mass of 7,500 tonnes, this will put a high challenge on present upload capacities (appr. 30 tonnes for the Shuttle) and makes also the target date rather optimistic. The travel package foresees a six-day tour, including two days of basic training and a 48 hours stay in the Hotel. Furthermore foreseen are:

- A platform area (for spacemen docking)
- A service area (solar panels, radiators and so on)

- An entertainment area.

(By courtesy of Shimitzu Corp.)

Fig. 9.3: Shimizu Space Hotel project (1989)

The entertainment element needs to be elaborated. Attracting people for a simple stay will not be possible, therefore specialists have developed a list of "Leisure Activities" which will be needed in such Space hotel in order to offer an acceptable package.

A list based upon Collins and Ashford [26] is presented in table 9.2. Note that the activities have been ordered in sequence of increased additional costs (for example gardens could be simultaneously used for Life System Support).

ACTIVITY	EXAMPLES	EQUIPMENT
Earth Observation	Land formations, weather phenomena, hometown...	Window, camera
Astronomical Observations	Planets, Stars, Nebulae	Telescope, camera
Low-gravity sport	Gymnastics, ball games	Equipped room (handrails, padded walls)
Low-gravity phenomena observation	Liquids, ballistics, magnetic effects, plants	Laboratory space and equipment
Low-gravity swimming	Partial immersion	"Swimming room"
Artificial-gravity swimming	Rotating water drum	"Water Drum"
Gardens	"Giant" plants, low-gravity adaptations	Large dedicate volume
Extra-vehicular (space walk)	Visit to other facilities, outside view	Space suits and safety devices

Table 9.2 Space Hotel leisure activities

In this context we can note that the previously mentioned surveys [16, 17, 18] have found a number of preferred activities during a space trip from the respondents.

If we express this in ranked order of preference we come to the following classification:

1. Look at the Earth
2. Make a space walk
3. Do astronomical observations
4. Perform microgravity sports/experiences
5. Perform microgravity experiments
6. Undergo the re-entry phase.

It is interesting to note that these preferences seem to be "culture-free" because the same ranking is, with only marginal differences, found in all countries examined.

Of interest from a marketing point of view, it was also found that the duration of the trip offered had no paramount importance (provided evidently it is in the order of days).

Hotel Berlin

In Europe, a different approach is proposed. Based upon the Columbus module, Europe's contribution to the ISS, a modular circular concept is proposed by connecting twelve such modules. The "Space Hotel Berlin" [27] could accommodate four tourists per module, in total a capacity of 48 tourists. A number of inner modules would foresee crew quarters and dining and entertainment areas. The total mass would be in the order of 600 tonnes, slightly higher than the ISS mass.

The interesting aspect of this design is the reuse of existing hardware designs, which would considerably reduce design costs and production time. Evidently, DASA, who produced the initial Columbus module, is highly interested in this concept. A building time of only 10 years is estimated, so that, after proven functioning of the Columbus module as part of ISS, operations well before the year 2020 seem possible.

The "WAT&G" Hotel

Another similar idea from the Architecture Company WAT&G is based upon reusing the Space Shuttle fuel tanks (which are presently dropped and burn in the atmosphere upon re-entry). The circular design, equally allowing for outside accommodation and inner recreational facilities, would be able to accommodate 100 guests. The hotel is designed to operate with the future X-33 VentureStar that is scheduled to replace the Shuttle fleet from 2007 onwards.

TOURISM PROJECTS ON THE MOON AND MARS.

The best known project in this area is the Lunar Hilton Hotel. The design is not only impressive, as can be seen from the artist impression in figure 9.4, but a lot of publicity was gained by Hilton in trying to legally patent the concept. The 5-star hotel is designed to accommodate not less than 5,000 persons and has dimensions in the order of 325 m height and 1500 m diameter with, in addition, its own lake and beach. This gave raise to a legal discussion on the applicability of the Outer Space Treaty text (1), which clearly did not forecast such discussion at the time they were established and signed. On the other hand the concept of "free access to all areas, stations, installations, equipment and space vehicles on the Moon and other celestial bodies", which has also been imbedded in the Treaty, weakens any property rights claim.

(published in www.resonance-pub.com/space.htm)

Fig. 9.4 Lunar Hotel Project from Hilton)

It looks evident that such projects are not possible to realise in the near future, probably not even in the first half of the 21st Century. Other preliminary designs of this nature are reported to be under consideration by the previously mentioned Shimizu Corporation and also a Moon Habitat project has been developed by the Russian Design Bureau of General Machine Building, which is composed of underground habitats in part. The latter designwork was stopped in the late 80's.

9.6   PRICE OF SPACE TOURISM

Only in an indicative way can we deduce some cost and price figures from the previous projects. Although calculation bases are different and difficult to compare, an attempt is made to summarise them in table 9.3.

Some figures have been estimated because as they were not clearly indicated in the various publications. Furthermore the "ticket" price assumptions in most publications are very optimistic (such as 100% occupation rates); specifically in the Space Hotel areas price estimates range from 50,000 to 250,000 $ per stay as a function of the assumptions taken.

PROJECT	INVESTMENT COST	"TICKET" PRICE
Spacebus	20 billion $	4,500 $ (1997)
Ascender	3.4 billion $	20,000 $ (1997)
Kanko-Maru	11.6 billion $	90,000 $ (1998)
X-33	10 billion $ (est.)	100,000 $ (1998)
Shimitzu Space Hotel	100 billion $ (est.)	150,000 $ (1990)
Space Hotel Berlin	50 billion $	184,000 $ (1998)
Stay on Mir	N/A	20,000,000 $ (2000)

Table 9.3: Indicated Prices

We can now consider two approaches for the pricing aspect:

- A cost model from the supply point of view
- A design to cost approach from the demand point of view.

COST SIMULATION MODEL

One ROI (Return On Investment) model could run as follows:

$$G \times P = (TC + MC + PE + FC + DC) \times (1 + R) \qquad \text{(Eq.1)}$$

Whereby:

G = Total of yearly number of Passengers
P = Price per ticket
TC = Transport Cost per year
MC = Maintenance Costs per Year
PE = Personnel Expenses
FC = Fixed Costs per year (Insurance, Administration, Ground facilities)
DC = Depreciation Cost per year
R = Margin for Risk and Profit (example 0.08).

Each element can be further expressed as:

TC = (LC/PC) x WT x (G + I)
MC = 0.15 x TCC (assuming 15 % yearly maintenance)
FC = 0.15 x TCC (assuming a 10% insurance cost, certainly higher in the initial phase)
DC = TCC / t

With:

LC = Cost per launch (example Shuttle 110 million $)
PC = Payload Capacity per launch (example: Shuttle 29.5 tonnes)
WT = Weight per passenger, including food and other supplies (typically 100 kg per person)
G = number of Guests per launch
I = number of Instructors per launch (including replacement crew for Space Hotels
TCC = Total Construction Cost (for Hotels: Assembly complete in orbit)
t = expected lifetime

It is evident from this model that TC, the transport cost per year, will be a determining factor. For instance if we make a very simple assessment by putting all costs equal to the transport costs; we could simplify Eq. 1 as

G x P = 2 x (LC/PC x W x (G + I)) (Eq. 2)

Or, with

W = 100
I = 0.2G (20% professional crew)

G x P = 2 x (LC/PC x 100 x 1.2G) (Eq. 3)

Resulting in

P = 240 x (LC/PC) (Eq. 4)

Eq. 4 would indicate that LC/PC needs to be in the range of 210 $/kg, to come to acceptable prices for a 50,000 $ ticket (i.e. only 1 % of the present Shuttle price). The previously mentioned Ascender project (see also figure 9.5), with an operational launch cost of 30,000 $ for one trip with 4 passengers (400 kg) reaches an LT/PC ratio of 75 $/kg and therefore seems to have sufficient margin to have a feasible project appraisal.

(By courtesy of Bristol Spaceplanes Ltd)

Fig. 9.5 : The Ascender project (artists impression)

Pearsall [28] performs a different approach, based upon macroeconomic data, constructing a price-demand curve for space holidays. Also confronted with the uncertainty of the transport cost he concludes realistically:

> *Given the above it is probably not worthwhile spending significant sums on detailed designs or feasibility studies until 2005, when the first generation of reusable launchers will be available and there will be a useful amount of experience from the International Space Station. Only then it will be possible to make realistic and realisable plans.*

Abitzsch (16) has combined the various demand curves and found this relation for the ticket price in function of the number of expected clients:

Ticket Price = $1.125 \times 10^6 \, (N)^{-0.25}$ (in 1994 $), whereby

N = the number of passengers per year.

This means that for example with 1000 passengers a yearly revenue of only 200 million $ can be expected, which is certainly insufficient to develop a system and expect an acceptable ROI concept. For 10,000 passengers the revenue would be 1125 million $ per year which also severely limits the investment potential. Important from these figures is that a marketing strategy can be deduced in terms of

- Which market segment will I consider?
- How many passengers do I need to transport per year?
- Which is the resulting concept?
- Which is the resulting investment budget?

9.7 PHYSICAL DISTRIBUTION AND TARGET GROUPS

Specifically during the first phase, one targets:

- a relatively small population
- a very world-wide population spread
- a group of relatively wealthy people.

It looks evident that such a target group will not do the research on their own but will have dedicated research executed on their behalf, probably using electronic databases. This reasoning must have been the same for the early providers who clearly choose for the Web as a prime distribution channel. As an example, Space Adventures Ltd. is offering a trip for 98,000 $ under

http://www.spacevoyages.com

It is interesting to note that the choice of the vehicle offered is still declared as open. In the list of potential candidates we find the "Ascender" and the "Rotor".

Also in Europe, a space hotel trip can be registered under

www.spacetours.de

In view of the longer stay, the price range is in this case 250,000 $. Registration on the "waiting list" is, however, only 50$...
It shall be noted here that both sites are also offering more readily available space-near experiences, such as

- training sessions in ZPK, the Cosmonaut Training Centre near Moscow
- possibilities to witness launches
- parabolic flights.

The latter aspect is the most realistic experience one can offer on Earth to simulate weightlessness. A specially equipped and reinforced wide-body airplane performs a number of parabolas. During the descent phase zero gravity can be simulated for short periods of some 25 seconds. This experience is used to train astronauts on how to move and work in weightlessness conditions. (However, the rapidly varying g-loads heavily stress the neurovestibular system; the airplane-ride is therefore also often called the "vomit-express" in professional circles).

In line with the general chapter on distribution it has to be noted that in none of the other cases such a high emphasis is put on the use of the Internet as a distribution medium. In fact, many publications are concurrently available via the different Web sites. Supported by large organisations in this area are the following sites:

In the U.S., the official Space Transportation Association Web site

http://www.spacetransportation.org

And the Japanese space tourism Web site

http://www.spacefuture.com

However, before full commercialisation can be envisaged, a considerable number of legal and regulatory aspects will need to be solved first before viable large scale business can emerge. Already the following list, which has been collected in the NASA/STA study [20] p.12, is impressive enough:

Near-term Regulatory Issues:

- Experimental flight regulations
- Spaceport regulations

- Waiver of liability
- Space traffic management.

Near-term policy issues:

- Use of government assets
- Privatisation of zero-gravity flights
- Authority to license re-entry vehicle.

Longer-term issues:

- Certification of commercial transport systems operations
- Property rights (claims registration)
- Environmental issues (overland supersonic flights)
- Orbital debris removal.

It can only be hoped that these elements are tackled in time by the legal experts in order to avoid a later bottleneck in the implementation once the financial and technical problems are solved, a possibility that is becoming more and more realistic.

9.8 PROMOTION FOR SPACE TOURISM

The above-mentioned Web sites are the medium of excellence to reach the younger "@-public", who are anyway the highest potential future customers. Besides the commercial information mentioned they provide interesting information and background references on Space Tourism. Furthermore, a growing number of enthusiasts are putting material on the Web.

A well informed source provides the DLR scientist Reichert, who maintains the site:

www.alltra.de

NASA is very active in the field of Internet distributed information. Even for "standard" situations, a number of WWW pages can be consulted, amongst others, on:

- Current (weather) conditions on Mars:

http://humbabe.arc.nasa.gov/

- Future Mars missions:

http://spaceflight.nasa.gov/mars/

Specifically this last site provides extremely interesting artist impressions of planned future Mars habitations.

A more "classical" approach are congresses or symposia associated with a broad press coverage. In March 1997, the first "International Symposium on Space Tourism" (ISST) took place in Bremen, Germany, sponsored by inter alia DASA, DLR and the British Interplanetary Society. Some 100 experts, some also from all the major Space Agencies, presented the state-of-the-art designs of potential projects and discussed strategies. Also due to the involvement of "star" guests like Buzz Aldrin, the event attracted a lot of interest from not only the specialised but also the general press inside and outside Germany. This has led to a complete revival of interest in Space Tourism.

The message found in most newspapers was that Space Tourism would become realistic for prices in the order of 50,000 $. The general public was at that point in time probably for the first time confronted with a price tag that was "reachable".

9.9 THE PHILOSOPHICAL DIMENSION OF SPACE TOURISM

Previous dimensions are important in view of the feasibility for Space Tourism projects. The fact that people are found to be willing to pay amounts in the order of several millions of dollars for a short trip in space can, however, only be explained by a deeper and philosophical dimension.

Authors have noted that a new generation of travellers is emerging, in the sense that the traveller has ceased to be a tourist and has become a searcher. Motivations include the discovering of oneself and psychological mobility. Many travel agencies and folders are still using the Product Mix from the past and are not adapting to the emerging Life Cycle of new tourism products.

According to Moutinho [29] p.17, travel motivation can be grouped in a number of categories, as represented in table 9.4.

MAIN MOTIVATION	SUBGROUPING
Educational and Cultural	- Observe people in other countries - See particular sights and monuments - Gain a better understanding - Attend special cultural events
Relaxation, Adventure & Pleasure	- Get away from every day routine - Seek new experiences - Have fun and a good time - Have some sort of romantic experience
Health and Recreation	- Rest and recover from strain - Practice sport and exercise
Ethnic and Family	- Visit "roots" - Visit relatives and friends - Spend time with family
Social and "Competitive"	- Able to talk about places - Being "fashionable" - Show what one can afford

Table 9.4: Tourist Motivation elements

According to Anderson [30] pp.17-18, the challenge involved in travelling is based on the exploring instinct labelled the "Ulysses factor" (cf., the hero of Homer's Odyssey):

It is the need for exploration and adventure, involving an exciting and even (according to the individual's perception) risky action. It is a physical and intellectual need related to knowledge and curiosity.

This "Ulysses factor" is probably the main drive for individuals to be receptive towards Space Tourism offers, hence the most important aspect in the Marketing Mix. In a certain analogy to our topic, research has been done for example on the motivation to participate in the river rafting trips in the Colorado River Basin. These types of activities are classified under "extraordinary experiences" in socio-cultural research. Such experiences were found to be evaluated by the participants as important for "harmony with nature", "community feeling and spirit" and "personal growth and renewal".

As the authors of the study conclude: [31] p.41,

> *River rafting is a unique recreational form, but its power lies in the romantic cultural scripts that evolve over the course of the experience – the opportunity to participate in rites of intensification and integration and the return to an everyday world "transformed".*

We can safely assume that similar motivations will be applicable to the early space tourist who is looking for an extraordinary experience to break with day-to-day rhythms of life. Virtually each astronaut coming back from a space trip has gained a number of impressions and keeps transmitting them to audiences:
- the colours and beauty of the Earth's surface
- the lack of visual borders and the global, not nationally divided, Earth
- but also ... the fragility of the thin atmospheric layer.

Ashford and Collins [32] summarise this idea as follows:

> *Space travel will not be a panacea for all the ills of the modern world, but it should help to make the Earth a better place. Many of the astronauts who have been privileged to visit space have declared that their perspective of the Earth and the universe changed as a result of the experience. They became more concerned for the safety of the planet than with national advantage. When millions of people visit space it is very likely that the way the peoples of the Earth view their planet will change for the better.*

Sally Ride, the first American woman in space and now a highly esteemed presidential advisor, expressed this as follows [33]:

> *In the 21st Century, travel will become more commonplace. As technologies mature, travel agencies will begin booking passages on commercial "spaceliners" and adventurous travellers will enjoy vacations in orbit around Earth. As they rise above Earth's atmosphere, these space travellers will experience the transforming view of their home planet's oceans and lands, wrapped in a thin cocoon of air, set against the velvety blackness of space. (...) As we enter the new millennium, no one really knows where our scientific and technological innovations will lead. But our past reminds us that what we dare to dream today often becomes reality tomorrow.*

9.10 MARKETING PLAN

As a consequence of previous observations, one can distinguish four phases, for each of them a different marketing plan and strategy is needed. (Adapted from Collins and Ashford [26]).

Phase 1: The "Pioneering" Phase:
Price per trip: 100,000 to 1,000,000 $

The customers will be wealthy individuals, looking for a new challenge (the "ultimate kick"). They will probably have low requirements on comfort, leisure facilities or a prolonged stay in orbit. They will be ready to take the risks associated to such trip and probably willing to give a "waiver" to the organiser in case of an accident. Most probably they will not leave the space vehicle.

Their motive is merely the "adventurous" one; close to the earlier mentioned "Ulysses motivation". The risk factor will rather attract them than be a hindrance for their decision.

Early Ascender proto-flights, even involving co-financing of such type of projects, could be their "target" vehicle (having probably two pilots in the early phase, together with test flight equipment and one passenger only).

Phase 2: The "Exclusivity" Phase:
Price per trip: around 50,000 $

The target group will be different and more driven by a "competitive" motivation, i.e. being the first in a certain circle. However, one could also assume that these type of flights would attract people of a more modest income category who would be willing to bring a special "offer" (such as the ones now flying the Concorde once in their lifetime).

This phase will require a more sophisticated marketing strategy because the customer will make a choice between various alternatives. Therefore one can assume that a certain degree of comfort and an "added value" would need to be offered in terms of (leisure) facilities. In addition to this, the customer will have a critical evaluation of the safety aspects of the product proposed and past performance. This would point to a sort of approach similar to the "Kanko-Maru" one described earlier in the text.

Phase 3: The "Mature" Phase:
Price per trip: 5,000 to 10,000 $

Such trip falls within reach of a broader population, even in the order of more than one million candidates per year according to the data in (16). Customers will have to be offered an acceptable "package", more than a trip. Leisure facilities will have to be on board or on the station being docked to.

In order to accommodate such a number of passengers a fleet of vehicles will be needed; in this phase probably many competitors will enter the market leading even to higher price competition and falling prices which will gradually lead to the next phase.

Phase 4: The "Market" Phase
Price per trip: less than 5,000 $

Competition will be at full strength, similar to present competition for trips from Europe to America, South Africa and Australia. Providers will have to come up with novelties in order to compete in the market. The more traditional Marketing Mix instruments will play a major role again (Price and Promotion).

Each of the different phases will require a different Marketing Plan or, rather, the Marketing Plan will have to evolve constantly with the Life Cycle stage.

One particular aspect to be mentioned here again is the safety aspect. It is evident that the public will react to an excessive high number of accidents in the early phase of space tourism, hence the apparent importance. On the other hand, the associated risk will attract a potential group of "clients", in a similar way as activities such as skydiving, bungee-jumping and free mountain climbing.

This high-risk behaviour has been studied in the case of for example skydiving. Research teaches us [34] that this risk factor strengthens motivation in the process, going from efficacy feelings over identity formation to even transcendent experience. There are evidently close parallelisms between this target group and the space tourism target group (which, however, will have additional boundary financial constraints to take into account besides the purely motivational ones.)

9.11 CONCLUSION

Let us first summarise the main findings.

- The Tourism Market is looking for new Products
- Space Tourism perfectly fits into this pattern
- The turnover of the market is large enough to allow development of such Products
- There are sufficient early potential clients, certainly in the 50,000 $ / trip category
- No essential new technologies need to be developed, existing know-how can be used.

From this we can conclude:

- Space Tourism will develop once the launch costs falls in the feasibility "band-width"
- Space Tourism will become a commercial activity
- New transport means will have to be adapted based upon available technology
- Development will follow different phases, each of them requiring an adapted marketing strategy.

From this perspective, Space tourism looks a prime candidate for a PPP approach, whereby the tourist industry will indicate the basic design parameters and Space Agencies will use the existing technology to develop a corresponding vehicle. The first steps will be taken jointly: industry will market the product and Agencies will ensure safety and further product development. In the maturity phase, industry will run the business on a commercial basis.

It cannot be ignored that one can see a number of parallels with the present launcher evolution in Europe. ESA and CNES developed the Ariane launchers and managed the qualification phase and once qualified, Arianespace successfully commercialised the exploitation. With the market changing, new launchers were reconceived (such as Ariane 5) and, based upon the updated requirements of Arianespace, developed and qualified by the Agencies. A similar process could be well imagined for Space Tourism, probably starting from a modest, dedicated spaceliner (such as the Ascender or equivalent) and gradually expanding to commercial spaceliners (such as the Spacebus). From that point onwards the way to Space Hotels will be open and probably financed by industry in view of the commercial potentialities.

At this point it has to be mentioned that not only "space-enthusiasts" see perspectives in this emerging market; also "The Economist" concluded already in 1997:

> *If safety can be significantly improved, the main commercial market for manned space flight may be tourism. A surprising number of people might be willing to pay a large sum in order to take the ultimate in holiday snapshots.*

REFERENCES CHAPTER 9:

1. UNITED NATIONS, *Treaty on Principles Governing the Activities of States in the Exploration and Use of Outer Space.* (Washington, 19 December 1967).
2. WASSENBERGH, H., The Art of Regulating International Air and Space Transportation. *XXIII Annals of Air and Space Law.* (McGill University, Montreal, 1998) pp.201-229.
3. KURENT, H., Tourism in the 1990's: Threats and Opportunities. *World Travel and Tourism Review.* Vol. 1 (1991) p. 78-82.
4. ISLEY, C., New Product Research in the Travel and Tourism Market. Is it Possible? In *Travel and Tourism in Transition.* (ESOMAR, Dublin, May 1991) pp.117-128.
5. STIPANUK D., Tourism and Technology: Interactions and Implications. *Tourism Management.* (August 1993), pp. 267-278.
6. SPACE WATCH, Prize Offered to Promote Space Tourism, *Space Watch,* (June 1996), p.39.
7. DIAMANDIS, P., the X PRIZE Competition, in Haskell, G. and Rycroft, M., *New Space Markets,* (Kluwer, Dordrecht, 1998).
8. TSIOLKOVSKY, K., *The Purpose of Space Exploration* (in Russian), (Kaluga, 1929).
9. ZUBRIN, R., *The Case for Mars* (Free Press, New York, 1966).
10. WITT, S. and MOUTINHO, L., *Tourism Marketing and Management Handbook.* (Prentice Hall, Hertfordshire, 1989).
11. DAVID, L., Advocates of Space Tourism Say Market Will Develop Slowly. *Space News* (July, 19th, 1999) p. 34.
12. NAISBITT, J., *Global Paradox* (Avon, New York, 1994)
13. MATSUMOTO, S. et al., Feasibility of Space Tourism. *Proc. of the 17th Symposium on Space Technology and Science.* (Tokyo, 1990) pp. 2301-2308.
14. COLLINS, P. et al., Commercial Implications of Market Research on Space Tourism. *Proc. of the 19th Symposium on Space technology and Science.* (Yokohama, May 1994).
15. DASCH, P., Space Tourism: Getting Beyond the Giggle Factor. *Ad Astra Magazine.* (March/April 1996). P.40.
16. ABITZSCH, S., Global Market Scenario of a Space Tourist Enterprise. *Paper presented at the International Symposium on Space Tourism.* (Bremen, March 20-22, 1997).
17. BARRETT, O., *An evaluation of the potential demand for Space Tourism within the United Kingdom,* (Bournemouth University, Dorset, England, 1999).
18. ABITZSCH, S., Prospects of Space Tourism, *Paper presented at the 9th European Aerospace Congress,* (Berlin, 15 May 1996).
19. BERGER, B., Billionaire Shops for Space Tourism Vehicle. *Space News,* (May 10th, 1999) p.6.
20. O'NEIL, D. et al., *General Public Space Travel and Tourism.* NP-1998-3-11-MSFC, (NASA, March 1998)
21. BERGER, B., MirCorp Seeks Partnership in Renovation of Ageing Station. *Space News* (4 May 2000), p.4.
22. MIRCORP, *Press Release,* available under www.mirstation.com (June 19, 2000).
23. ASHFORD, D., A Development Strategy for Space Tourism. *Journal of the British Interplanetary Society.* Vol. 50 (1997), pp. 59-66.
24. ASHFORD, D., Orbital and Sub-orbital Passenger Transport. *Paper presented at International Symposium on Space Tourism.* (Bremen, March 1997).
25. COLLINS, P., The Japanese Rocket Society's Space Tourism Research. *Paper presented at International Symposium on Space Tourism.* (Bremen, March 1997).

26. COLLINS, P. and ASHFORD, D., Potential Economic Implications of the Development of Space Tourism. *Proc. of the 37th IAF Congress*, IAA-86-446 (Innsbruck, October 1986).
27. REICHERT, M., *The Future of Space Tourism.* ESA Report WPP-151 (ESA, Noordwijk, 1999).
28. PEARSALL, J., Space Hotels. *Journal of the British Interplanetary Society.* Vol. 50 (1997), pp. 67-80.
29. MOUTINHO, L., Consumer Behaviour in Tourism. *European Journal of Marketing.* 21(10), (1987) pp.5-43.
30. ANDERSON, J., *The Ulysses Factor* (Harcourt, New York, 1970).
31. ARNOULD, E. and PRICE, L., River Magic: Extraordinary Experience and the Extended Service Encounter. *Journal of Consumer Research*, Vol.20 (June 1993) pp.24-45.
32. ASHFORD, D. and COLLINS, P., *Your Spaceflight Manual. How you could become a Tourist in Space within Twenty Years.* (Eddison Sad, London, 1990), p.114.
33. RIDE, S., Today's Dreams, Tomorrow's Realities: Science in the New Millennium. In *The World Almanac* 1999. (Primedia, New Jersey, 1999) p.36.
34. CELSI, R., ROSE, R. and LEIGH, T., An Exploration of High-Risk Leisure Consumption through Skydiving. *Journal of Consumer Research*, Vol.20 (June 1993) pp.24-45.

LIST OF ACRONYMS

AF:	Award Fee
AFC:	Administrative and Finance Committee
ASE:	Association of Space Explorers (astronauts and cosmonauts)
ASI:	Agenzia Spaziale Italiana (Italian National Space Agency)
ATV:	Automated Transfer Vehicle (to ISS)
CAD:	Computer Aided Design
CAE:	Computer Assisted Engineering
CER:	Cost Estimation Relationship
CIAC	Canadian ISS Access Company (planned)
CIS:	Commonwealth of Independent States (ex-USSR)
CNES:	Centre Nationale d'Etudes Spatiales (French Space Agency)
CNSA:	Chinese National Space Administration
CPFF:	Cost-Plus-Fixed-Fee
CPIF:	Cost-Plus-Incentive-Fee
CPPF:	Cost-Plus-Percentage-Fee
CSA:	Canadian Space Agency
CTC:	Cost to Completion (overall all-in cost)
DARA:	Deutsche Agentur für Raumfahrtangelegenheiten (now DLR)
DLR:	Deutsche Forschungsanstalt für Luft- und Raumfahrt (German National Space Agency)
DoD:	Department of Defense (of the U.S.)
DTC:	Design to Cost
EC:	European Commission
ECOS:	ESA's Costing Software
ECSS:	European Cooperation for Space Standardisation
ECU:	European Currency Unit (Predecessor of EURO)
ERA:	European Robotics Arm
ESA:	European Space Agency

ESTEC:	European Space and Technology Centre (of ESA)
EU:	European Union
EURO:	European common currency
EVA:	Extra Vehicular Activity ("Spacewalk")
FFP:	Firm-Fixed-Price
FPI:	Fixed-Price-Incentive
FY:	Fiscal Year (U.S. budget appropriations)
GEO:	Geostationary Orbit (35,780 km)
GNP:	Gross National Product (key indicator of a countries wealth)
GNSS:	Global Navigation Satellite System
GPS:	Global Positioning System (Navigation)
IGA:	Inter-Governmental Agreement (on ISS)
IGS:	Interconnection Ground Subnetwork (ESA Intranet)
INCPO:	International Classification of Nonprofit Organizations
IMC:	Integrated Marketing Communications
IPR:	Intellectual Property Rules
ISPR:	International Standard Payload Rack
ISRO:	Indian Space and Research Organisation
ISS:	International Space Station
LCC:	Life Cycle Cost (Total cost of all project phases)
LEO:	Low Earth Orbit (600 – 2,000 km)
MEO:	Medium Earth Orbit (5,000 – 20,000 km)
MEURO:	Million Euro
MIPS:	MIR Interface to Payload Systems
MOU:	Memorandum of Understanding
MTBF:	Mean Time Between Failures (reliability indicator)
MTBR:	Mean Time Between Repairs (reliability indicator)
MUSC:	Microgravity User Support Centre (USOC at DLR)

limited partnership · 29
publicly owned debt and equity · 29
FOCUS · 274
FUEL CELLS · 161, 284

G

GAGARIN, YURI · 19, 106, **227**, 252
GAGARIN COSMONAUT TRAINING CENTRE 306
GALILEO · **35**, 36, **37**, 38, 242
  benefits · **39**
  cost estimates · **38**
  financing · **38**
  Strategy considerations · **36**
  system architecture · **36**
GALLUP SURVEY · **98**, 206
GIOTTO · 126, 199
GLENN, JOHN · **296**
GLOBALISATION · 17, **28**
GLONASS SYSTEM · **36**, 37
GOLDIN, DAN · 86, 87
GPS · 35, **36**, 37, 39, 40
GTS · 274

H

HARSH-ENVIRONMENT INITIATIVE · 170
HELIOS · 7
HORIZON 2000 · 86
HOTOL · 307
HUBBLE TELESCOPE REPAIR · 146
HUMAN EXPLORATION · 229

I

IMC · *See* integrated marketing communications
IMPACT · 171
IN SEARCH OF EXCELLENCE · 84
incentives
  award fees · **134**
  cost incentives · **134**, 135
  delivery incentives · **134**, **135**, 136
  performance incentives · **134**, 136
  systematic considerations · **135**

INCPO · 330
INDIA · 6, 205, 247
INDUSTRIAL EVOLUTION · 17
INDUSTRIALISATION · **276**, 268, 270
INFORMATION CHANNEL COMPARISON · 167
INFRASTRUCTURE · 3
INNOVATION 164
  Chain-link model · **166**
  process · 151, 162, 166
INSURANCE 145
  in orbit performance · 144
  launch and commissioning · 144
  pre-launch and construction · 144
  premiums · **80**, 144, 145
INTEGRATED MARKETING COMMUNICATION 190, 220
  windows 95 · **191**, 192
INTELLECTUAL PROPERTY 177
  data base protection · 178
  data protection · **177**, 178, 267
  and PPP · **181**
  and WTO · **180**
  and TRIPS · **180**
  copyright protection · 178
INTERCONNECTION GROUND SUBNETWORK (IGS) · **155**, 281
INTER-GOVERNMENTAL AGREEMENT (IGA) · **283**
INTERNATIONAL SPACE STATION · *See* ISS
INTERNET 154, **176**
  and Mars Pathfinder · **174**
  generated revenue · 173
ISPR · **261**, 279, 280
ISS · **259**
  and philosophy · **223**, 240, 242
  and promotion · **187**, 188, 189
  cost per phase · **268**
  data distribution · **282**
  development cost · **278**
  main characteristics · **260**
  pricing policy · **279**
  sponsoring approach · **286**
ISS AND INTERCULTURAL RELATIONS · 292
ISS INDUSTRIAL APPLICATIONS · 274
ISS USER CENTRE · 282

J

JAPAN · 6, 7, 16, 22

K

KANKO-MARU · 309, 315, 324
KENNEDY, JOHN F. · **19**, 120
KNOWLEDGE POOL · 100

L

LANGUAGE PROBLEMS · **214**
LAUNCHER PERFORMANCE · 145
LAUNCHING STATE · **180**
LCC · *See* Life Cycle Cost
LEAD TIME TO PRODUCTION · **159**
LECTURE FROM SPACE · **287**
LICENSE · **109**, 183
LIFE CYCLE COST · 126
   commitment · **128**
LIFE SCIENCE RESEARCH · **105**
LINGUISTIC PROBLEMS IN COMMUNICATIONS · **207**
LONG-TERM SPACE POLICY COMMITTEE · **200**
LUNAR HILTON HOTEL · **313**

M

MANAGEMENT BY OBJECTIVES · **88**
MARKET GROWTH · **3**
MARKET MYOPIA · 19
MARKET PULL · 84, **166**
MARKET VOLUME · **2**, 3
   European market share · **13**
   global market · **2**, 3
   public market · **2**
   space turnover · **3**
MARKET VOLUME FOR TOURISM · **302**
MARKETING
   4Ps approach · **64**, 72
   criticism · **60**, 241
   customer oriented · **47**, 197
   definition · **47**
   historical context · **46**
   philosophy · **67**, 241
   place · **65**, 153
   price · **65**, 113, 162, 278
   product · **65**, 75
   production oriented · **46**
   promotion · **67**
   sales oriented · **46**
MARKETING MIX   45, 64
   definition · **48**
MARKETING MIX FOR NONPROFIT ORGANISATIONS **69**, 71
MARS · 158, 206, 259
MARS OBSERVER · 90
MARS/EARTH CHARACTERISTICS · **301**
MASS MEDIA ADVERTISING · 191
MEDIA CHOICE · **209**
MEDICAL COUNTERMEASURES · 105
MEDICAL SPIN-OFF · **105**
MERCURY PROJECT · 120
MERGERS · 24
MICROGRAVITY EXPERIMENTAL CONDITIONS · **279**
MICROGRAVITY EXPERIMENTS COST · 278
MILESTONES IN SPACE EXPLORATION · 227
MIR · 32, 155, 214, 258, 281
MIRCORP · **306**, 307
MTBF · **127**
MTBR · **127**
MULTIMEDIA TECHNIQUES IN COMMUNICATIONS · **211**
MULTIPLIER FACTOR · **103**

N

NASA · 120, 168, 169, 174, 175, 279, 281
NASA'S MIDLIFE CRISIS · **86**
NAVIGATION APPLICATIONS · **37**
NETWORK OF TECHNICAL CENTRES · **15**
NEW PRODUCT DEVELOPMENT · 83
NONPROFIT MARKETING · **49**
NONPROFIT SECTOR
   classification · **53**
   definition · **51**
   differences with profit sector · **56**
   macroeconomic importance · **55**
   organisations · 53

NONPROFIT SECTOR · **51**
NUTRITION · **106**

O

OBERTH, HERMANN · 240
OFFICE FOR COMMERCIALISATION · **264**
OFFICE OF ADVANCED CONCEPTS · **169**
OPERATIONS PLANNING · 157
ORBITAL TOURISM FLIGHTS · 298
ORTHOSTATIC INTOLERANCE · **106**
OSTEOPOROSIS · **106**, 275
OUTER SPACE TREATY · **178**, 238, 297
OUTSOURCING · **129**
OVERRUNS · 63, **118**

P

PAID PUBLICITY · **207**
PARADIGM SHIFT · **226**
PARAMETRIC ESTIMATING · **122**
PATENTS **181**
    and Bayh-Dole act · **182**
    and ESA · **170**
PERFORMANCE INDICATORS · **90**
PLASTIC INDUSTRY AND ESA · 170
POTENTIAL ISS APPLICATIONS · **272**
PPP · *See* Public-private Partnership
PRICE DISCRIMINATION · **114**
PRIVATISATION · **268**
PRODUCT LIFE CYCLE · **158**
PRODUCTION FLOW · **117**
PRODUCTIVITY · **88**
PROGRESS ROCKET · 91
PROMOTION **187**, 192
    Advertising · **187**, 188, 190
    and NASA · 267
    in Europe · 192
    personal selling · **188**
    public relations · **188**
PUBLIC ACCESS TO SPACE · **296**, 297, 305, 307
PUBLIC EXPENDITURE · **22**
PUBLIC SERVICE ADVERTISEMENTS · 210, **215**
PUBLICITY EXPENDITURE PER MEDIUM · **208**
PUBLIC-PRIVATE PARTNERSHIP · 15, **33**

principles · 33

Q

QUALITY · **95**
    of new products · **100**
    public response · **97**

R

R&D · 34, 165
    and marketing conflicts · **162**
    characteristics · **165**
    effectiveness index · **85**
    expenditure per sector · **168**
    worldwide spending · **162**
RACE TO THE MOON · **228**
RADIUS PROJECT · **170**
RAPID RAMP-UP TECHNIQUES · 158
RECOVERY EFFORT · **94**
REHABILITATION · **106**
RELIABILITY **91**
    launch failures · 91
RETURN ON INVESTMENT · 87, 190, 265, 314
RISK COMMUNICATION · **93**
RISK MANAGEMENT · **113**, 122, 137, 140
    flow · **140**
    policy · 140
    risk assessment · 139, 140
    risk monitoring · 140
ROGERS COMMISSION · 92
ROI · *See* Return on investments
RONIN · **233**, 234
ROUND TRIP PROPAGATION DELAY · **157**
RUSSIA · 6, 25, 27, 212

S

S*T*A*R*S · **287**
SACRIFICES IN PATTERNS OR BEHAVIOUR · 114
SAGAN · **239**
SAGAN ARGUMENT · **290**
SAGE · **274**
SAPHIR · **236**

SAPPHO · 83
SATELLITE INTERNET MARKET · 4
SEA LAUNCH · 24, 25, 26
   consortium · 26
SECURITY AND PEACEKEEPING · 201
SEGMENTALISTIC STRUCTURES · 230
SERVICE MARKETING · 48, 49
SERVQUAL · 95
   questionnaire · 95
SHAPE AND SHARE MESSAGE · **204**, 220
SHUTTLE · 176, 296, 312
SHUTTLE TRANSPORTATION SYSTEM · 120
SMALL AND MEDIUM ENTERPRISES · 161
SOCIAL MARKETING · 65
SOCRATES · 242
SOR-MODEL · 97
SOYUZ ROCKET · 99, 306
SPACE BURIALS · 12
SPACE COMPANIES · 25
SPACE EXPENDITURE · 6
   agency part · 7
   evolution · 8
   in comparison with R&D · 7
   in function of GNP · 6
SPACE EXPLORATION · 80
SPACE FRONTIER · 11, **223**, 224
SPACE HOTELS · 299
   hotel berlin · **311**, 312, 314
   leisure activities · 311
   Shimizu hotel · 309
SPACE INDUSTRY TURNOVER · 10
SPACE MILESTONES · 228
SPACE OBJECTS · 32
SPACE PRODUCT MARKET · 81
   budget sharing markets · 82
   international markets · **81**, 82
SPACE PRODUCTS 76
   direct product · **76, 78**
   in nonprofit marketing · 76, 77, 197
   non-tangible products · 78
SPACE RACE · 1, 227
SPACE STATIONS · **259**, 261
   historical overview · 258
SPACE SUPPORT IN U.S. · 288
SPACE TOURISM
   and NASA · 305
   cost simulation model · 314
   demand curves · 316
   indicated prices · 314
   market · 303
   marketing plan · **322**, 323, 324
   regulatory issues · 318
SPACE TRANSPORTATION ASSOCIATION · **305**, 318
SPACEBORNE EARLY WARNING SYSTEMS · 202
SPACEBUS · **308**, 314, 325
SPACECRAFT FAILURES · 78
SPACEHAB · 85, 143, 287
SPACELAB · 82, 155, 261, 271
SPIN-IN · 167
SPIN-OFF 101
   effect · **102**, 104
   examples · 104
   process · 101
SPONSORING · 32
SPOT · 7
STRATEGIC ALLIANCES · 15
STS · *See* Shuttle Transportation System
SUBORBITAL TOURISM FLIGHTS · 299
SYSTEM INTEGRATION AND NETWORKING (SIN) · 166

T

TECHNOLOGY
   performance evolution · 229
   and marketing conflicts · 162
TECHNOLOGY LIFE CYCLE (TLC) · **151**, 159
TECHNOLOGY PUSH · 84
TECHNOLOGY PUSH-MODEL · 166
TECHNOLOGY TRANSFER SYSTEM · 168
TEILHARD DE CHARDIN · 238
TELEMEDICINE 108
   advantages · 108
   obstacles · 109
TELESCIENCE · 155, 157
TERRAFORMING · 301
TIPPLER, FRANK · 237
TLC · **151**, 158, 159, 167
   Advanced stage · 160
   and marketing · 159
   Cutting Edge · **151**, 159, 160
   decline · **160**, 161
   mainstream · 160
   maturity · 160
   State of the art · 159

TOURISM ON MOON OR MARS · **312**
TOURIST MOTIVATION ELEMENTS · **321**
TQM · **79**
    rules according to Deming · **79**
TRACKING AND DATA RELAY SATELLITES · **155**
TRADE-OFFS · **129**
TRANS-UTILITARIAN RATIONALE · **236**
TRENDS · **10**
    in Europe · **12**
TSIOLKOVSKY, KONSTANTIN · 240, **241**

U

ULYSSES FACTOR · **321**
USE OF SPACE DATA AND ASSETS · 3
USER RELATIONS · 85
USER SUPPORT CENTERS 169
    MUSC · **169**
UTILISATION SCENARIO OF ISS · **279**

V

VENTURE CAPITAL · 29, **30**, 31
VOLCANIC ERUPTIONS · **202**
VON BRAUN, WERNHER · 19, **223**, 299
VOYAGER CAPSULE · **252**

W

WITCH DOCTORS · 76
WORLD SPACE AGENCY · **17**, 18
WTO · 28, **180**

X

X PRIZE FOUNDATION · **298**
X-33 REUSABLE CONCEPT · **307**
XENOPSYCHOLOGY · **246**, 252